THE NEW GROVE
TURN OF THE CENTURY MASTERS

THE NEW GROVE
DICTIONARY OF MUSIC AND MUSICIANS

Editor: Stanley Sadie

The Composer Biography Series

BACH FAMILY

BEETHOVEN

HANDEL

HAYDN

HIGH RENAISSANCE MASTERS

ITALIAN BAROQUE MASTERS

MASTERS OF ITALIAN OPERA

MODERN MASTERS

MOZART

SCHUBERT

SECOND VIENNESE SCHOOL

TURN OF THE CENTURY MASTERS

WAGNER

THE NEW GROVE®

Turn of the Century Masters

JANÁČEK   MAHLER
STRAUSS   SIBELIUS

John Tyrrell
Paul Banks
Donald Mitchell
Michael Kennedy
Robert Bailey
Robert Layton

MACMILLAN

18,832   RO9198
780·92 TYR

First published in
The New Grove Dictionary of Music and Musicians®,
edited by Stanley Sadie, 1980

The New Grove and *The New Grove Dictionary of Music and Musicians*
are registered trademarks of Macmillan Publishers Limited, London

First published in UK in paperback with additions 1985 by
PAPERMAC
a division of Macmillan Publishers Limited
London and Basingstoke

First published in UK in hardback with additions 1985 by
MACMILLAN LONDON LIMITED
4 Little Essex Street London WC2R 3LF
and Basingstoke

The New Grove turn of the century masters.—
(The New Grove composer biography series)
1. Composers—Europe   2. Music—Europe
—History and criticism—19th century
3. Music—Europe—History and criticism
—20th century
I. Tyrrell, John   II. Series
780′.92′2   ML240.4

ISBN 0–333–38541–1 (hardback)
ISBN 0–333–38542–X (paperback)

First American edition in book form with additions 1985 by
W.W. NORTON & COMPANY
New York and London

ISBN 0-393-01694-3 (hardback)
ISBN 0-393-30098-6 (paperback)

Printed in Great Britain by
Redwood Burn Limited, Trowbridge, Wiltshire,
and bound by Pegasus Bookbinding, Melksham, Wiltshire.

Contents

List of illustrations

Illustration acknowledgments

We are grateful to the following for permission to reproduce illustrative material: Moravské Muzeum, Brno (figs.1–3); Österreichische Nationalbibliothek, Vienna (figs.5, 9, 10, 12, 14); Gesellschaft der Musikfreunde, Vienna (fig.8); Bayerische Staatsbibliothek (Handschriftenabteilung), Munich (fig.13); BBC Hulton Picture Library,

General abbreviations

A	alto, contralto [voice]	к	Köchel catalogue [Mozart]
a	alto [instrument]	Ky	Kyrie
acc.	accompaniment, accompanied by	lib	libretto
addl	additional		
addn	addition	Mez	mezzo-soprano
Ag	Agnus Dei	movt	movement
appx	appendix		
		n.d.	no date of publication
B	bass [voice]		
b	bass [instrument]	ob	oboe
Bar	baritone [voice]	orch	orchestra, orchestral
bar	baritone [instrument]	orchd	orchestrated (by)
bn	bassoon	org	organ
		ov.	overture
c	circa [about]		
cl	clarinet	perc.	percussion
conc.	concerto	perf.	performance, performed (by)
Cr	Credo		
Cz.	Czech	pic	piccolo
		pubd	published
db	double bass	pubn	publication
dbn	double bassoon		
		qnt	quintet
edn.	edition	qt	quartet
facs.	facsimile	R	photographic reprint
fl	flute	rearr.	rearranged (by/for)
frag.	fragment	reorchd	reorchestrated (by)
		repr.	reprinted
gui	guitar	rev.	revision, revised (by/for)
		Russ.	Russian
HMV	His Master's Voice		
hn	horn	S	soprano [voice]
hpd	harpsichord	str	string(s)
		Swed.	Swedish
inc.	incomplete	sym.	symphony, symphonic
Jb	Jahrbuch [yearbook]	T	tenor [voice]
Jg.	Jahrgang [year of publication/volume]	t	tenor [instrument]
		timp.	timpani

tpt	trumpet	v, vv	voice, voices
transcr.	transcription, transcribed (by/for)	va	viola
		vc	cello
trbn	trombone	vn	violin

U.	University

Symbols for the library sources of works, printed in *italic*, correspond to those used in *Répertoire international des sources musicales*, Ser. A.

Bibliographical abbreviations

AMf	*Archiv für Musikforschung*
AMw	*Archiv für Musikwissenschaft*
BMw	*Beiträge zur Musikwissenschaft*
ČSHS	*Československý hudebni slovník*
DJbM	*Deutsches Jahrbuch der Musikwissenschaft*
GfMKB	*Gesellschaft für Musikforschung Kongressbericht*
HR	*Hudebni revue*
HRo	*Hudebni rozhledy*
HV	*Hudebni věda*
JbMP	*Jahrbuch der Musikbibliothek Peters*
Mf	*Die Musikforschung*
MJb	*Mozart-Jahrbuch des Zentralinstituts für Mozartforschung* [1950–]
ML	*Music and Letters*
MMC	*Miscellanea musicologica* [Czechoslovakia]
MMR	*The Monthly Musical Record*
MQ	*The Musical Quarterly*
MR	*The Music Review*
MT	*The Musical Times*
MZ	*Muzikološki zbornik*
NRMI	*Nuova rivista musicale italiana*
NZM	*Neue Zeitschrift für Musik*
OM	*Opus musicum*
ÖMz	*Österreichische Musikzeitschrift*
PNM	*Perspectives of New Music*
PRMA	*Proceedings of the Royal Musical Association*
ReM	*La revue musicale*
RIM	*Rivista italiana di musicologia*
RMI	*Rivista musicale italiana*

SH	Slovenská hudba
SM	Studia musicologica Academiae scientiarum hungaricae
Smv	Suomen musiikin vuosikirja
SMw	Studien zur Musikwissenschaft
SMz	Schweizerische Musikzeitung/Revue musicale suisse
STMf	Svensk tidskrift för musikforskning
ZfM	Zeitschrift für Musik
ZIMG	Zeitschrift der Internationalen Musik-Gesellschaft
ZMw	Zeitschrift für Musikwissenschaft

Preface

This volume is one of a series of short biographies derived from *The New Grove Dictionary of Music and Musicians* (London, 1980). In its original form, the texts were written in the mid-1970s, and finalized at the end of that decade. For this reprint, they have been re-read and modified by the original authors and corrections and changes have been made. In particular, the bibliographies have been expanded and brought up to date to incorporate the findings of recent research.

The fact that the texts of the books in this series originated as dictionary articles inevitably gives them a character somewhat different from that of books conceived as such. They are designed, first of all, to accommodate a very great deal of information in a manner that makes reference quick and easy. Their first concern is with fact rather than opinion, and this leads to a larger than usual proportion of the texts being devoted to biography than to critical discussion. The nature of a reference work gives it a particular obligation to convey received knowledge and to treat of composers' lives and works in an encyclopedic fashion, with proper acknowledgment of sources and due care to reflect different standpoints, rather than to embody imaginative or speculative writing about a composer's character or his music. It is hoped that the comprehensive work-lists and extended bibliographies, indicative of the origins of the books in a reference work, will be valuable to the reader who is eager for full and accurate reference information and who may not have ready access to *The New Grove Dictionary* or who may prefer to have it in this more compact form.

S.S.

LEOŠ JANÁČEK

John Tyrrell

CHAPTER ONE

Early life and studies, 1854–80

Leoš Janáček was born in Hukvaldy, northern Moravia, on 3 July 1854 and christened Leo Eugen the next day. 'Leo' persisted in official documents during his student days but had already given way informally to a more Slavonic 'Lev' ('lion') with which he signed his first surviving letter (to his uncle, Jan Janáček, 26 May 1869) and under which some of his earliest works were published. 'Lev' itself gave way to another Slavonic variant, 'Leoš', which Janáček first used in January 1880, in his correspondence with his fiancée, after a Christmas reunion had also encouraged his use of the intimate 'thou'.

Janáček's family was of the Czech *kantor* tradition. Both his grandfather (Jiří, 1778–1848) and his father (Jiří, 1815–66) were teachers, musicians and leading cultural figures in the poor communities they served. In 1838 Janáček's father married Amálie Grulichová and in 1848 he moved with her and their five children to a full teacher's post in the village of Hukvaldy. Leoš was the fifth of the nine children born there and to relieve the crowded home he was sent, when he was 11, to be a chorister at the Augustinian 'Queen's' Monastery in Old Brno. The choir school (founded in 1648 as a conservatory) was then in decline. The Austro-Prussian War of 1866 disrupted it and when it reassembled the effects of the Cecilian movement began to be felt: instrumental teaching ceased and the choristers became a purely

1

vocal ensemble. Brno nevertheless played a vital role in Janáček's development; in particular the choirmaster of the monastery, Moravia's leading composer, Pavel Křížkovský, took a keen interest in his musical education.

Janáček was to follow his family's teaching tradition and by September 1869, after completing his basic schooling, including three years at the German college in Old Brno, he went on a state scholarship to the Czech Teachers' Institute (c.k. Slovanský Ústav ku Vzdělání Učitelů). He passed his final examinations (excelling in music, history and geography) in July 1872 and served the compulsory two-year period of unpaid teaching at a school run by the institute. In 1872 he also took over the monastery choir when Křížkovský was transferred to Olomouc Cathedral. Janáček's hard, thorough work enabled him to perform a wide variety of music at the services – Palestrina, Lassus, Haydn and contemporary Czech and German works – and led to his appointment (1873) as choirmaster of a working-men's choral society, Svatopluk (founded 1868). Janáček raised the level of the society from its *Liedertafel* traditions, moving the concerts out of the taverns into the new Beseda Hall, and widened the repertory. It was for Svatopluk that he wrote his first secular compositions, simple four-part settings of folk texts in the style of Křížkovský.

In the autumn of 1874, after completing his period of unpaid teaching, Janáček obtained leave to study under Skuherský at the Prague Organ School (a type of conservatory that placed special emphasis on the training of church musicians). During the year that Janáček was there he completed the first two years of the three-

1. Leoš Janáček and Zdenka Schulzová before their marriage in 1881

year course. Extremely poor, with no money even for a piano, he was unable to take full advantage of Prague's musical life. Several student exercises, mainly church and organ works, date from this period. Returning to Brno in 1875 he resumed all his previous activities: his teaching, conducting the monastery choir (for which he wrote an offertory, *Exaudi Deus*, his first work to be published, in 1877) and Svatopluk, from which, however, he resigned in October 1876, soon after becoming conductor (1876–88) of the Czech middle-class Beseda choral society (founded 1860). After a few months he turned the male-voice Beseda choir into a mixed body, and, with help from the monastery choir and pupils from the institute, he mustered a force of 250 singers for large-scale choral works, Mozart's Requiem (1877) and Beethoven's *Missa solemnis* (1879). He also championed Dvořák, introducing to Brno audiences his Moravian Duets and the Serenade for Strings, a work which probably stimulated the composition of Janáček's own works for string orchestra, the Suite (1877) and the Idyll (1878). He came to know Dvořák personally and the two men went on a walking tour of Bohemia in the summer of 1877. Janáček's string works show that he had not entirely digested the influences he had encountered (severe modal writing encouraged by the Cecilian movement, Dvořák, and Wagner's *Lohengrin*), nor, as can be seen in the awkward part-writing and clumsy modulations, had he achieved complete technical proficiency.

Although Janáček returned to Prague (June–July 1877) for a month of special study, he finally decided on further studies abroad and, on paid leave from the Teachers' Institute, he enrolled at the Leipzig Conservatory (Oct 1879–March 1880). There his teachers included Oskar Paul (whose history lectures he also attended at the university) and Leo Grill; though initially attracted to Paul, Janáček settled down to a course of extremely hard study under the strict and systematic Grill. Again poverty prevented his taking full advantage of his surroundings: he attended the Gewandhaus concerts but never went to the opera. Under Grill his interest in composition developed; his last Leipzig composition, the Schumannesque *Zdenčiny variace* ('Zdenka variations') for piano, show a smoother and more imaginative technique than his earlier works. With the renewal of his leave Janáček was able to complete the year abroad. Plans to continue in Paris under Saint-Saëns came to nothing, as had an earlier plan to study in St Petersburg under Rubinstein, and instead he went to Vienna in April 1880. He was, however, no happier at the Vienna Conservatory than in Leipzig, partly because he considered his teacher Franz Krenn not sufficiently strict and perhaps because he was anxiously trying to fit too much into a short time. Among other works (most of them now lost), he wrote a violin sonata, a string quartet and a song cycle, *Frühlingslieder*, which he entered unsuccessfully in a competition.

CHAPTER TWO

1881–1904

Before leaving for Leipzig, Janáček had already become informally engaged to his piano pupil Zdenka Schulzová (1865–1938), who was the daughter of Emilian Schulz, the director of the Teachers' Institute in Brno. Janáček had chronicled his unhappy, isolated life in almost daily letters to her; one reason for moving to Vienna was to make visits to her more frequent. On his return to Brno they were officially engaged and, with Janáček recognized by the education ministry as a 'full teacher of music' (May 1880) at the Teachers' Institute, they were married on 13 July 1881, shortly before Zdenka's 16th birthday. In addition to all his earlier activities Janáček began to realize his ambition of founding an organ school in Brno. A committee was established under the auspices of the Jednota pro Zvelebení Cirkevní Hudby na Moravě (Society for the Promotion of Church Music in Moravia) and on 7 December 1881 Janáček was appointed director; teaching began in September 1882, at first in the Teachers' Institute. From 1886 to 1902 he also taught music at the Old Brno Gymnasium. At the Beseda he added to the repertory some of Dvořák's major choral works as well as works by Brahms, Smetana, Tchaikovsky, Saint-Saëns and Liszt; he established singing and violin classes (1882), a permanent orchestra and (1888) piano classes. When the Provisional Czech Theatre opened in Brno in 1884 he saw the necessity for a journal to review its activities. This was the *Hudební listy*, published by the Beseda; Janáček was editor and a chief contributor. The journal lasted until 1888, though Janáček's uneasy relationship with the conservative Beseda continued until 1890. His married life, too, was no easier. The tensions between a fervently patriotic Czech and a very young, unimaginative girl from a German middle-class background proved unbearable and the couple separated from 1882 (soon after the birth of their daughter Olga) until 1884. A son, Vladimír, was born in 1888 but died of scarlet fever in 1890. Disillusioned by Leipzig and Vienna Janáček had stopped composing; in 1884 he resumed and then wrote the *Mužské sbory* ('Male-voice choruses') and the chorus for mixed voices, *Kačena divoká* ('The wild duck'). The latter was for school use; Janáček dedicated the former to Dvořák, who was startled by the boldness of the modulations.

Three years after the opening of the new Brno theatre, Janáček began to compose his first opera *Šárka*, to a verse libretto by Julius Zeyer based on Czech mythology. Zeyer had intended the work for Dvořák (who toyed with it) and consequently refused the unknown and inexperienced Janáček permission to use his text. By then Janáček had already written and revised the work; undaunted, he went on to score two of the three acts. It remained unperformed until 1925. While working on *Šárka* he was invited by František Bartoš, a fellow teacher at the Old Brno Gymnasium, to help him collect folksongs in northern Moravia (1888). This was Janáček's first visit to his native region since childhood and its impact on him was decisive. Hurt no doubt by his disappointment over *Šárka*, he turned his back abruptly on its somewhat gauche Romanticism and, for a few years, immersed himself completely in Moravian folk music. In addition to the folksong editions he brought out with Bartoš, he popularized his discoveries in a series of orchestral dances and dance suites, the *Lašské tance* ('Dances from Lašsko'), the Suite for orchestra, the folk ballet *Rákos Rákoczy*, hurriedly put together for the 1891 Jubilee Exhibition in Prague, and even the one-act opera *Počátek románu* ('The beginning of a romance', 1891) which consists of little more than folkdances with added voice parts. The libretto was adapted from a short story by Gabriela Preissová, who wrote the play *Její pastorkyňa* ('Her foster-daughter'). It was probably when Janáček realized the far greater possibilities of this play as the basis for a serious opera, also in a Moravian folk setting, that he became dissatisfied with his unassuming and favourably received earlier opera. Withdrawing it after four performances (1894) he set to work on *Jenůfa* (as the work has become known outside Czechoslovakia).

The long period of composition of *Jenůfa* (1894–1903) cannot be explained merely by Janáček's other activities, which by this time consisted of his teaching posts at the Teachers' Institute, the Old Brno Gymnasium and the Organ School, his folksong editions with Bartoš, and his work on the committee of the Prague Ethnographic Exhibition (1895), nor by the illness and subsequent death at almost 21 of his only surviving child, Olga, which clouded over the final stages of composition. There was a large interval, possibly of five years, between the composition of the first act of *Jenůfa* and the rest of the opera, during which much of Janáček's approach to opera and to composition seems to have been rethought. The earliest indications of this were in the cantata *Amarus* (c1897), the first work in which his distinctive idiom was apparent. During this period he began to abandon the number opera; he also integrated folksong fully into his music so that while its influence can be discerned in certain rhythmic and melodic formations, it ceased to overshadow his own musical personality. In about 1897 he began to formulate a theory of 'speech-melody', which was to influence his approach to the voice line and indeed his whole musical idiom for the rest of his life.

Jenůfa was thus a very different work from its predecessor. The success of its première in Brno (21 January 1904) was however probably due more to its Moravian setting than to the provincial audience's awareness of its stature. The performances suffered from a tiny and inadequate orchestra and Janáček, moreover, made substantial alterations before the work was published (1908). He had submitted the scores of both *The Beginning of a Romance* and *Jenůfa* to the Prague National Opera before settling for Brno premières. Karel Kovařovic, chief conductor at Prague, eventually came to see *Jenůfa* at Brno but still declined to take it up; possibly he remembered Janáček's scathing criticism of his own opera *Ženichové* ('The bridegrooms') many years earlier (1887).

CHAPTER THREE

1904–16

In 1904, the year of the Brno première of *Jenůfa*, Janáček reached the age of 50 and resigned from his post at
the Teachers' Institute. He concentrated on composition
and on running the Organ School, which with a bigger
grant acquired a new building in 1908 to accommodate
70 students. Janáček remained director until 1919, re-
fusing a surprising offer in 1904 of the directorship of
the Warsaw Conservatory.

Janáček's next opera *Osud* ('Fate', 1903–7) is a curi-
ous, semi-autobiographical work. Because of its reflection of contemporary life (it includes scenes at a spa and
at a music academy) and modish international influences
(Charpentier and Puccini) Janáček hoped for a production of it in Prague, this time not at the National
Theatre, but at the newly opened theatre in the
Vinohrady district. Although the work was accepted
there, production was continually postponed and despite Janáček's threatening lawsuits it was never performed during his lifetime. *Fate*'s exploration of unusual
subject matter – a hallmark of Janáček's later operas –
and a widening of the musical language mark an
advance on *Jenůfa* but its clumsy, home-made libretto
(by Janáček and a 20-year-old schoolteacher) has stood
in its way. This can be said, too, of his next opera, *Výlet
pana Broučka do měsíce* ('Mr Brouček's excursion to
the moon'), which Janáček, after working to little effect
with a long succession of 'librettists', adapted himself

from Svatopluk Čech's satirical novel. Both operas were preceded by a period of casting around for suitable librettos: before *Fate* he considered Merhaut's novel *Andělská sonata* ('The angel sonata'); before *Brouček* he considered plays by other Czech and Moravian composer–writers such as Ladislav Stroupežnický, Q. M. Vyskočil, the Mrštík brothers and again Gabriela Preissová. He also made sketches for a setting of Tolstoy's *Anna Karenina*.

Janáček began *Brouček* in 1908 and wrote it in spurts, with many revisions, over the next ten years. Most of his other compositions during this period were chamber and vocal works. For the piano he wrote two cycles, a series of miniatures that recalled his childhood in Hukvaldy, *Po zarostlém chodníčku* ('On the overgrown path', 1901–8) and four more substantial pieces *V mlhách* ('In the mists', 1912). Together with the two surviving movements of a piano sonata (1905), written under the impact of the death of a workman killed in an anti-German demonstration, these constitute virtually all Janáček's mature solo piano music. The piano sonata had been first performed at the Klub Přátel Umění (Friends of Art Club), founded in 1900 and whose music section Janáček founded in 1904. The possibility of performances of chamber works there led to the composition of the *Pohádka* ('Fairy tale', 1910) for cello and piano, a piano trio (1908–9), now lost, based on Tolstoy's *Kreutzer Sonata* (Janáček allegedly incorporated material from it into his string quartet with the same sub-title), and the arrangement of *Lidová nokturna* ('Folk nocturnes', 1906) for two-part female chorus and piano. The club also sponsored the first published vocal score of *Jenůfa* (1908).

The finest works from this period are the three great

11

male-voice choruses written for Ferdinand Vach's Moravian Teachers' Choir, *Kantor Halfar* (1906), *Maryčka Magdónova* (1906–7) and *Sedmdesát tisíc* ('The 70,000', 1909). These are the culmination of the line of choruses that Janáček had written continuously from his first compositions for Svatopluk and antedate the great creative period of his last 12 years almost by a decade. The texts, drawn from Petr Bezruč's *Slezské písně* ('Silesian songs'), had a deep social and patriotic appeal for Janáček and perhaps this fact, together with their setting near his native Hukvaldy, explains the strong response they elicited from him.

By his 60th birthday, shortly before the outbreak of World War I, Janáček was respected in Brno as a composer and as the director of the chief music teaching institution in Moravia; outside Moravia he was almost unknown. He was at least more prosperous: in 1910 he moved into a house specially built for him in the grounds of the Organ School; in 1912 he and his wife took a holiday abroad on the Adriatic coast of present-day Yugoslavia. He had begun to write non-operatic works for larger forces, such as the tone poem *Šumařovo dítě* ('The fiddler's child', 1912) and the cantatas *Na Soláni Čarták* ('Čarták on the Soláň', 1911) and *Věčné evangelium* ('The eternal gospel', 1914). The war had had its effect on his work. When Vach's male choir was disbanded during the war, he formed the Moravian Women Teachers' Choir, for which Janáček wrote his women's choruses, *Vlčí stopa* ('The wolf's trail'), *Hradčanské písničky* ('Songs of Hradčany') and *Kašpar Rucký*, all completed in 1916. His pro-Russian sympathies during the war stimulated the composition of a three-movement orchestral work, *Taras Bulba*

(1915–18), based on Gogol's novel. And as the war gave rise to the possibility of an independent Czech nation, he embarked on the writing of a sequel to *Mr Brouček's Excursion to the Moon, Výlet pana Broučka do XV. století* ('Mr Brouček's excursion to the 15th century'), where, as in Čech's novel, the Prague landlord finds himself in the midst of the patriotic Hussite wars. Janáček's aim here was not satirical, as in the first excursion, but an appeal to a more heroic past, and it was to the first president of the Czechoslovak republic, T. G. Masaryk, that the score was dedicated. After the declaration of the republic in 1918, one of Janáček's first works, programmatically linked to the idea of the Czech nation and regeneration, was the symphonic poem *Balada blanická* ('The ballad of Blaník', 1920).

This activity would scarcely have come about without one major factor: the acceptance in 1915 of *Jenůfa* by the Prague National Opera, after the persistent intervention of Janáček's friends Dr František Veselý, his wife Marie Calma Veselá and the critic and librettist Karel Šípek. This unexpected outcome persuaded Janáček to make a final revision of *Mr Brouček's Excursion to the Moon*, which he did with the help of F. S. Procházka (poet of the *Songs of Hradčany*) and a newspaper editor, Viktor Dyk. The first part was completed by an epilogue showing Brouček back home. Janáček wrote four versions of the epilogue before he finally discarded it when, with F. S. Procházka as his librettist, he added the sequel *Mr Brouček's Excursion to the 15th Century* to make a one-evening 'bilogy'.

The Prague première of *Jenůfa* took place on 26 May 1916 under Karel Kovařovic who, as a condition for accepting it, stipulated that he be allowed to revise it. It is likely that in addition to his careful rehearsal of the work his smoothing out of Janáček's idiosyncratic orchestration contributed to the work's instant and sustained success. Universal Edition arranged to take over the rights of the score abroad and promoted productions in Vienna, Berlin and other German cities, and the Prague German writer, Max Brod, who had reviewed the opera enthusiastically, made the German translation. Brod became Janáček's champion, the translator of all his subsequent operas except *Brouček*, and his first biographer.

CHAPTER FOUR

The last years, 1916–28

In the years between the Brno and Prague premières of *Jenůfa* Janáček was not composing at full capacity. His chief interest was opera and there was little chance of his operas being performed outside Brno and thus little incentive for further works. *Jenůfa*'s success in Prague and abroad changed this. The amazing creative upsurge in a man who was well into his 60s can also be partly explained by his patriotic pride in the newly acquired independence of his country and perhaps most of all by his friendship with Kamila Stösslová, 38 years younger than himself and the wife of an antique dealer in Písek, Bohemia, from whom Janáček had received provisions during the war. Janáček identified the heroines of his female-dominated operas with Kamila. The first work inspired in this way was the song cycle *Zápisník zmizelého* ('The diary of one who disappeared', 1917–19), a setting of a series of poems purporting to be written by a rich farmer's son to explain his attraction to a gypsy and his decision to desert his own family for her. While Janáček never went as far as deserting his wife for Kamila (though he considered this in the year before his death), he did write continual, almost daily, letters to her and, in the last year of his life, kept a special Kamila diary. His attentions were received with little warmth or understanding, though this did nothing to diminish his ardour or his creative zest: from 1919 to 1925 he composed three of his finest operas, *Kát'a Kabanová* (1919–21), *Příhody Lišky Bystroušky* ('The cunning little vixen', literally 'The adventures of the vixen Bystrouška', 1921–3) and *Věc Makropulos* ('The Makropulos affair', 1923–5). In between he found time to compose incidental but by no means insubstantial works. The First String Quartet was written in a matter of days in 1923. The same year he began a four-movement symphonic work, *Dunaj* ('The Danube'); he wrote the wind sextet *Mládí* ('Youth') in the spring of 1924 and the Concertino for piano and chamber ensemble in the spring of 1925.

By the time he had passed his 70th birthday Janáček's change in fortune was remarkable. He had retired from the Brno Organ School (which in 1919 joined the Beseda music school to become the Brno Conservatory) and instead gave composition master classes (in Brno) for the Prague Conservatory (1920–25). Though *Brouček* had only grudgingly been produced in Prague in 1920 (the only Janáček operatic première outside Brno), all his new operas were taken up immediately by Brno, followed by Prague (*Kát'a Kabanová*: Brno, 1921, Prague, 1922; *The Cunning Little Vixen*: Brno, 1924, Prague, 1925; *The Makropulos Affair*: Brno, 1926, Prague, 1928). Even the early *Šárka* had been remembered and was performed in Brno in 1925. Universal Edition published each new opera as it came out. In 1924 *Jenůfa* received important premières at Berlin (under Kleiber) and at the Metropolitan, New York. *The Cunning Little Vixen* had been presented as part of the ISCM Festival in Prague in the spring of 1925; in September the chamber section of the festival took place in Venice and Janáček went there to hear his First Quartet given a very warm reception. One of the honours that marked his 70th birthday was a doctorate from the Masaryk University in Brno (28 January 1925), a distinction he never ceased to cherish, signing correspondence and all his compositions 'Dr Ph. Leoš Janáček'.

Early in 1926 Janáček wrote his largest purely orchestral work, the five-movement Sinfonietta, commissioned (as a 'fanfare') for the Sokol gymnastic festival and dedicated to Rosa Newmarch. Shortly after he had completed the work he went to England for a week at the invitation of a Committee organized by Mrs Newmarch, with whom he corresponded and who had enthusiastically taken up his cause. His visit coincided with the General Strike, but a concert, which included most of his chamber works to date, took place as planned, at the Wigmore Hall (6 May). In the next months he wrote the Capriccio for piano left hand and chamber ensemble, the *Glagolská mše* ('Glagolitic Mass') and extended to 18 the eight movements of his *Říkadla* ('Nursery rhymes') written in 1925.

His fame continued to grow. While *Jenůfa* was performed in scores of German opera houses, *Kát'a Kabanová* began to penetrate into Germany with performances in Cologne (1922, under Klemperer) and Berlin (1926). His native Hukvaldy unveiled a plaque in July 1926, and on 10 February 1927 he was elected, together with Schoenberg and Hindemith, a member of the Prussian Academy of Arts. The Sinfonietta began to be widely known; Klemperer conducted performances in 1927 in Berlin and New York. The *Glagolitic Mass*, which received its Brno première in December 1927, was performed in Berlin in 1928. Meanwhile Janáček began work on his last opera, *Z mrtvého domu* ('From the house of the dead'). This was to occupy him from early 1927 to his death, though he broke off in 1928 (29 January–19 February) to write the Second String Quartet *'Listy důvěrné'* ('Intimate letters') and later that year he began, somewhat reluctantly, to sketch the incidental music for Gustav Hartung's production in Berlin of Hauptmann's *Schluk und Jau.*

In 1921, the year he began *The Cunning Little Vixen*, Janáček had bought a cottage in Hukvaldy, to which in 1924 and 1925 he added land from the adjacent forest. Like Luhačovice and, in later years, Slovakia, this became a favourite holiday place, but while the visits to Luhačovice, a fashionable spa where he went for his rheumatism, were also social events, Janáček retreated to Hukvaldy to do the concentrated creative work that his increasingly public life in Brno made difficult. That he had been born there made the Hukvaldy ties especially strong. In July 1928, after a short treatment in Luhačovice, he went to Hukvaldy and was joined for the first time there by Kamila, her 11-year-old son and, for the first few days, her husband. Janáček took with him the sketches of *The Danube*, *Schluk und Jau* and the fair copy of the third act of *From the House of the Dead* to revise. While searching for Kamila's son, who had got lost on one of their expeditions, Janáček caught a chill, which rapidly developed into pneumonia. On 10 August he was taken to the nearest large town, Moravská Ostrava, where he died at 10 a.m. on Sunday 12 August. His funeral, held in Brno on 15 August, was a large public event at which the final scene of *The Cunning Little Vixen* was played. Shortly after his death

2. *Leoš Janáček outside his house in the grounds of the Brno Organ School (20 June 1928)*

his Second String Quartet was given publicly (Janáček had been present at private performances); in April 1930 *From the House of the Dead* received its première in a version prepared by Janáček's pupils Břetislav Bakala and Osvald Chlubna. Zdenka Janáčková died ten years after her husband, in 1938, leaving her estate to the Masaryk University to form the basis for the Janáček Archive, now in the Music History Division of the Moravian Museum; Kamila Stösslová died in 1935.

CHAPTER FIVE

Range and background

Janáček worked in only two genres continuously during his career: opera and the unaccompanied (mostly male-voice) chorus. From *Šárka* to the end of his life there was scarcely any time in which he was not writing, revising or at least planning an opera. Writing choruses served as a preparation and as a substitute for operas, and as soon as his opera writing began in earnest after *Jenůfa*'s acceptance in Prague his interest in choruses waned. Apart from a few patriotic *pièces d'occasion* the only chorus after 1916 was *Potulný šílenec* ('The wandering madman', 1922) with words by Tagore, whose visit to Prague in 1921 had made a great impression on Janáček. It is significant that the most vital and important choruses, the Bezruč group, were written in the years of his greatest despondency over his operatic career. Janáček set these choruses in a more obviously dramatic manner than he had used previously, with indirect speech turned into direct speech and solo voices to suggest individual characters, a technique he developed memorably in two later works. In *The Wandering Madman* he added a solo soprano to the male-voice chorus to sing the questions of the young boy, while in *The Diary of One who Disappeared* he turned a song cycle for solo tenor into a chamber cantata by including a mezzo-soprano to sing the gypsy's words and an offstage chorus of three solo female voices.

Large-scale cantata works, on the other hand, do not seem to have provided Janáček with a compensatory non-operatic outlet. Despite the promising start made in *Amarus* (the first work to show the new directions in which he was moving during the long gestation of *Jenůfa*), the two cantatas from his dispirited middle period, *Čarták on the Soláň* (1911) and *The Eternal Gospel* (1914), are not among his most memorable works; their lack of direction suggests the uncertainties that beset him at this time. It was only in his last years that he wrote such characteristic though very different works such as the *Glagolitic Mass* (1926) and *Nursery Rhymes* (1925–7).

After his student works Janáček seems to have made no systematic attempt to write instrumental works which would have secured performances more easily than operas. He wrote chamber and solo instrumental works reasonably frequently after 1900, mostly for local performance at the Friends of Art Club; but apart from a burst of orchestral arrangements of folk-dances in the early 1890s (in connection with his efforts to promote Moravian folk music) he turned to large-scale orchestral composition only in his 60s. Although operas took precedence in his creative upsurge after 1916 and non-operatic works were written only while he paused between operas or versions of one opera, the confidence and creative zest in all his late instrumental works made them far superior to any he had so far produced. In particular the final version (1921) of the Violin Sonata, the two string quartets (1923, 1928), the wind sextet *Youth* (1924), the two chamber piano concertos (Concertino, 1925; Capriccio, 1926) and the Sinfonietta (1926) represent Janáček's finest instrumental works.

Almost all of Janáček's works, whether vocal or instrumental, had a programmatic origin, although purely musical considerations generally predominated once the piece had been set in motion. Later the programmatic beginnings may seem no more than a curiosity, as in the original programmes of the Concertino and the Sinfonietta (the first, a set of animal scenes; the second, glimpses of Janáček's Brno), neither of which are essential for understanding these works. Many of his later works look back to his childhood, for instance *Youth* and *On the Overgrown Path*. Others have more enigmatic titles. The most plausible explanation of *In the Mists* is that it is Janáček's comment on his uncertain career before the Prague *Jenůfa*. His strongest creative impulse, however, was erotic; it is possible to see in the succession of female heroines a commentary on his attitude to Kamila Stösslová – from wishful thinking in *Kát'a* and the *Vixen* (the neglected wife who takes a lover; the vixen as fulfilled wife and mother) to sober reality in *Makropulos* (the fascinating but unmoved Emilia Marty, 'cold as ice'). In his last opera he turned his back tersely on female stage characters, though he identified Kamila with Akulina, the tragically murdered wife in Šiškov's tale. Both his powerful string quartets also had erotic origins. The First Quartet was inspired by Tolstoy's tale of marital unhappiness and infidelity (*The Kreutzer Sonata*) and his Second Quartet was to have been not merely 'intimate letters' to Kamila but 'love-letters'.

Before the advent of Kamila, Janáček's operas were

both less powerfully charged in this way and less successful, the only exception being *Jenůfa*, with its two
superb female portraits. The two operas directly after
Jenůfa were built round men, the satirical antihero of
the Prague landlord Mr Brouček, and in *Fate* the composer Živný, in whom Janáček saw himself. It is significant that the most effective section of his first opera,
Šárka, is the point where Šárka's passionate hatred for
Ctirad and men gives way involuntarily to love.

Although most of Janáček's operas are concerned
with death, they are equally concerned with renewal. He
was attracted to plots like those in *Jenůfa* or the *Vixen*
which passed beyond tragedy to regeneration. Something of this optimism, coupled with the grit and determination that made Janáček persevere until he won
recognition in his 60s, is evident in the sub-title of the
Capriccio, 'Vzdor' ('Defiance'), and in the nature of the
piece. Although Janáček was not attracted to the piano
he agreed to write a concerto for Otakar Hollman, who
had lost his right hand in the war, and responded to this
challenge not with the imaginative technique of Ravel or
the wit of Prokofiev but with an equally defiant orchestration which pitted the pianist's left hand against an
ungainly ensemble of three trombones, tuba, two trumpets and flute/piccolo. Old age and fame, however, made
Janáček unbend and relax; these years saw the affection-
ate wind suite *Youth* and the cheerful *Nursery Rhymes*.

Janáček's ardent nationalism also affected his choice
of subject matter. His strong reaction to the long
Austrian domination of his country is evident from the
Slavonic (Czech, Moravian, Russian etc) background to
almost all his works. Even his *Glagolitic Mass* sets a
church Slavonic text ('Glagolitic' refers in fact not to

the text, but to its original script, an early form of Cyrillic). His pan-Slavonicism took a particularly active form in his espousal of Russian culture. He knew and admired much of Russia's literature and music (Tchaikovsky's operas were a strong influence) and taught himself Russian from the age of 20. In 1897 he helped found a Russian club in Brno, serving as chairman from 1909 to its enforced disbanding in 1915 and again after the war (1919–21). He visited Russia in 1896 and in 1902, when he sent his daughter to St Petersburg to study. Both his children were given Russian names; the dedication of *Jenůfa* to his daughter Olga was written in Cyrillic characters. Russian authors, including Ostrovsky, Tolstoy, Gogol and Dostoyevsky, inspired some of his greatest works.

It has been convenient in the country of his birth to find evidence of a social awareness in Janáček, for instance in the Bezruč choruses or in his 1905 piano sonata inspired by the death of a worker in an anti-German demonstration. In these cases it is difficult to disentangle purely nationalistic reactions from social conscience. In the Austrian-ruled country in which he lived most of his life Czech speakers were the social and economic underdogs and Janáček took their part with typical fervour. In later years his nationalism became muted: earlier he had boycotted German trams and the German opera house; after 1916 Janáček was happy to entrust the fate of his operas to the Viennese Universal Edition and did all he could to encourage the translation of his operas into German and their production on Austrian and German stages.

CHAPTER SIX

Style

Although Janáček was in fact born before the last great
wave of Romantic composers – Mahler, Wolf, Sibelius,
Strauss and Reger – his most characteristic music was
written at the end of his life, in the 1920s, and belongs in
sound and spirit with the music of the younger generation around him. This is not to deny that his musical
language was grounded in the 19th century. Despite
some modal tendencies from Moravian folk music and
the whole-tone inflected passages which began to appear
in his music after his flirtation with French music
around 1900, his harmony operates functionally, and
the dissonance which had grown from the characteristic
major 9th formations in *Jenůfa* and *Fate* into the
increasingly harsh combinations in *From the House of
the Dead* reinforces rather than negates the tonal frame-
work. Janáček's treatment of tonality was generally
instinctive; the tonal plan of a piece is often more the
result of his gravitating towards his favourite keys
(Db/C♯, Ab) than of large-scale planning. In the mature
works a piece seldom ends in the key it began. Although
key signatures linger on in the instrumental works up to
the First String Quartet (in his operas Janáček abandoned them as early as *Brouček*) their use becomes
increasingly haphazard. Some of the later examples – in
the *Diary* or *Taras Bulba* – are for very short passages
in works basically without key signatures where the

music has slipped into A♭ and, as if to show he was on home ground, Janáček signalled the fact with a temporary key signature of four flats.

Janáček's music often contrasts types of harmony. In opera this procedure is particularly useful in underlining characterization; for example, Kát'a's music stands out

Ex.1 *Kát'a Kabanová* Act I

Con moto (♪ = 112)

KABANICHA

Ty bys mo-hla ml-čet,   mo-hla ml-čet.

Nik-do   se   tě nep-tal.   Ne - za - stá - vej se ho.

Katěrina: You are like a mother to me. How can you think otherwise?
            And Tichon loves you in the same way.
Kabanicha: You ought to keep quiet! No one asked you to speak! Don't
            stick up for him.

against Kabanicha's harsher, more dissonant idiom (see
ex.1). The lushness of Kát'a's music, typical of Janáček's
poignant 'dolce' manner, is not so much one of rife
chromaticism (though chromatic alteration has a part in
it) but an intensity built up through added chords and
appoggiatura formations (creating a type of melody that
often seems to overshoot its mark) and by the gentle
tension of 6-4 chords. Other harmonic contrasts with a
diatonic norm that Janáček uses are whole-tone inflected
harmony and harmony constructed from 4ths, a procedure that gives rise to many of his typical melodic pat-

Ex.2 String quartet no.1, 1st movt

terns made up of 4ths, 5ths and 2nds (see ex.2). The use of
harmonically contrasting blocks to build up a musical
structure is found equally in his non-operatic music, as
with the 'normal' version (ex.3*a*) of the Adagio tune in
the Second String Quartet, third movement, compared
with the more dissonant and constricted version that
follows it (ex.3*b*) and with the jubilant, wide-spaced
major version (ex.3*c*).

Another type of contrast that Janáček's music exploits is one of conflicting elements, as for instance in
the last movement of the Violin Sonata, where a tiny
repetitive fragment on the violin interrupts the would-be
broad-arched tune of the piano. Sometimes such 'inter-
ruption motifs' are repeated to form a disruptive ostinato, as in the overture to *Makropulos*, where a high
degree of tension is generated by the precarious balance
of melodic foreground and disruptive background. The
tension of these rapid ostinato figures is increased by
their generally jagged outlines with awkward jumps.

29

Ex.3 String quartet no.2, 3rd movt

Style

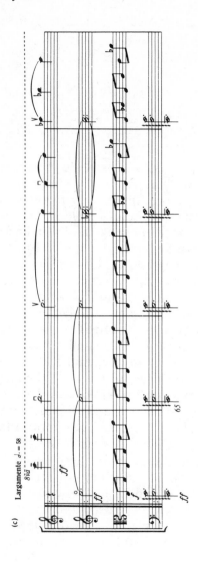

31

Other characteristic rhythmic procedures derive from
folksong. Janáček sometimes used Moravian mirror
rhythms (see ex.4); the trick of immediately repeating a

Ex.4 Sinfonietta, 4th movt

tune in a rhythmically displaced version is also based on
folk precedents. One trait that is evident throughout
Janáček's music, even before his involvement with
Moravian folk music, is the shortness of melodic breath,
despite his attempts, under Leipzig and Viennese
influence in his earliest works, and later under French
and Italian influence, to lengthen it. In fact much of
the awkward gait and stumbling harmonic rhythm of
Janáček's earliest music would seem to arise from
forcing his utterance into unsuitable foreign moulds.
Czechoslovakia looks both west towards Germany and
Austria and east towards Russia and Poland, a fact that
is evident from both occidental and oriental features in
its folk music. Janáček's Moravian affinity with the East
is very clear when he is compared to his Bohemian contemporary, Fibich, whose Western outlook is matched
by his fluent command of the tonal and developmental
techniques of Western music. Janáček's music progresses
by repetition and juxtaposition; accordingly much of
its formal organization consists of piling up repetitive
blocks (as in the Sinfonietta) and in a variety of rondo
forms. The few sonata forms he used depend on melodic
contrast rather than on tonal tension. He had little sympathy with another feature of Austro-German tradition,

counterpoint, and after his student works what little counterpoint there is seems to be the result of a montage of added or staggered parts, for instance in the complex textures of the late choruses.

Janáček's handling of voice parts is very demanding. Most characteristic are his mercilessly high tenor parts, both for soloists and chorus. His preference for high solo voices is clear from the *Glagolitic Mass*, where the soprano and tenor have most of the important solos, while the contralto makes her first appearance as late as the Sanctus ('Svet'). It is typical that in *Jenůfa* the four main solo parts are given to two tenors and two sopranos. Janáček used lower female voices less to depict older women than to suggest provocative eroticism (Zefka in the *Diary*, Varvara in *Kát'a*, the prostitute in *From the House of the Dead*) or for travesty roles (in *From the House of the Dead* and the *Vixen*). It is in the latter opera, with the complications of its animal world, that he specified children's voices, a resource he had previously resisted for the juvenile roles in *Jenůfa* and *Fate*.

With its addition of voices for specific roles and its instructions for dramatic lighting, the *Diary* shares the dramatizing instincts of the late choruses, but Janáček nevertheless observed the metrical and rhyme schemes of its poems to such an extent that the voice parts sound much more structured and conventionally melodic than those of his operas from the same period. In the *Nursery Rhymes* he even made a feature of the obvious sing-song verse, employing for it a deliberately non-realistic vocal style with two or three voices to each part. The *Glagolitic Mass*, despite its important solos, depends greatly on its chorus parts, but here Janáček was thinking in subjective, almost operatic terms: in an introduction to the work, he wrote that the tenor was a 'high priest', the soprano an 'angel' and the chorus 'the Czech people'. Another feature of these vocal works is the importance of the instrumental part. Both *Nursery Rhymes* and the mass are flanked by instrumental movements (the mass ends with two – an organ voluntary and an orchestral 'intrada'). At the heart of the Creed ('Věruju') in the mass there is a long orchestral commentary on the Incarnation and Crucifixion; the 13th piece of the *Diary*, where the young man loses his virginity, is a piano solo. Most of the *Nursery Rhymes* have an instrumental rather than a vocal climax; the voice parts often seem merely to provide programmes for the pieces, set in a terse, even perfunctory manner.

Janáček's orchestration is one of the most distinctive aspects of his style, though one that took time to develop. His earliest attempts, as in the *Dances from Lašsko*, are unexceptionable, overfull perhaps, with no evidence of a particularly acute ear. The experienced Kovařovic felt he had to reorchestrate *Jenůfa* not so much because the score was idiosyncratic but because he thought it clumsy and impractical in the theatre. While Janáček's first two operas were written in vocal score and then orchestrated (the scoring of the third act of *Šárka* was even undertaken by a pupil), from *Fate* onwards Janáček wrote straight into full score, a procedure that bespeaks both a greater confidence and the greater importance of the orchestral sound in the initial inspiration. (It is not clear what happened in *Jenůfa* since Janáček's sketches and autograph score have not survived.) By *Kát'a* he had evolved his characteristic sound: although capable of great sweetness there is a roughness caused by the unblended layers of orchestra and by the awkward writing in individual parts. In time he increasingly abandoned the middle ground for the extremes; the first sketches for *From the House of the Dead* reveal how often he thought instinctively in terms of three low trombones and three high piccolos, a sound image only partly modified in later versions.

Janáček's orchestration, like most aspects of his music, often has a programmatic origin, for example many of the uses of unusual instruments (the xylophone in *Jenůfa* and *Makropulos*, the sleigh bells in *Kát'a*). Dostoyevsky's *From the House of the Dead* provided a wealth of 'natural' sounds (chains rattling, anvil and hammer strokes, whiplashes and saws) which found their way into the final score; several characters are associated with individual instruments. But despite the array of percussive instruments, Janáček's last opera shares with some of the late instrumental music a frequently spare texture. One reason for this is that with this opera he went over to drawing his own staves, a habit he had adopted many years earlier in his non-operatic works. Although his instrumental writing remained difficult to the end of his life, with awkward figurations (especially in the rapid ostinatos), passages in extreme registers (e.g. very high violin parts) and seemingly unidiomatic writing, he became accurate and economical with his instrumental effects. The bizarre combinations in the Capriccio and the Sinfonietta are well calculated, and the colouristic devices in the Second String Quartet (e.g. the extensive use of *sul ponticello*) are all vividly effective in performance.

CHAPTER SEVEN

Operas

A striking feature of Janáček's mature operas is his successful use of a wide range of material that seems at first operatically intractable. His earliest operas inevitably follow the path of Czech nationalist opera. *Šárka*, based on events of the 'Amazon' war in Bohemia's mythic history, continues the story of Smetana's *Libuše*; *The Beginning of a Romance* owes much to Dvořák's early village-based comedies; even *Jenůfa*, often considered unique in portraying the serious side of the Moravian village, was anticipated by J. B. Foerster's *Eva* (1897), similarly based on a play by Preissová. But thereafter Janáček's operas luxuriate in unconventional and varied subject matter. *Fate* presents contemporary scenes from a composer's life: at a fashionable spa, at home (a lively scene ending in the simultaneous demise of the composer's wife and his mother-in-law) and at the conservatory. Mr Brouček, a late 19th-century philistine, travels uncomprehendingly from his native Prague to the moon (inhabited by aesthetes) and then to the fervidly patriotic times of the Hussite wars. *The Cunning Little Vixen* is based on a tale woven round a series of 200 drawings illustrating the adventures of a vixen and published as a popular serial in a newspaper. *The Makropulos Affair* is about a 300-year-old woman whose strange story emerges only with the unravelling of a complicated legal case. *From the House of the Dead* sets Dostoyevsky's prison memoirs (lightly disguised as fiction) – a remarkably modern-sounding reportage with almost no narrative thrust. Only *Kát'a Kabanová*, a tale of adultery on the banks of the Volga, has a more conventionally operatic libretto, and Janáček's setting of Ostrovsky's play is one of at least seven.

Janáček was his own librettist for all the play-based operas (*Jenůfa*, *Kát'a* and *Makropulos*; *Šárka* was intended as an opera libretto from the outset) and provided serviceable librettos chiefly by ruthlessly condensing the originals. But for *The Beginning of a Romance* (based on a short story), *Fate* (Janáček's own scenario) and *Brouček* (two satirical novels), he used other librettists, though taking a steadily increasing part himself. His frustration in failing to find a satisfactory partner for *Mr Brouček's Excursion to the Moon* resulted in his writing his own librettos for the *Vixen* and *From the House of the Dead*. By the end of his life Janáček was so at home with this side of opera writing that his 'libretto' for *From the House of the Dead* consisted merely of a few sheets providing an order of events with page-references to his copy of Dostoyevsky. He then worked directly from the novel, translating it from the Russian as he composed.

The casts of the play-based operas were reasonably small (though this did not prevent Janáček from jettisoning some of the smaller characters) but the potential casts of the novel-based ones were huge and Janáček did his best in *Brouček* and the *Vixen* to cut them down by combining them, as he did with his bold amalgam of Ostrovsky's Kuligin and Kudrjáš in *Kát'a*. The *Brouček* combinations – basically the earth characters transformed into different guises on the moon and

3. *A page from the autograph score of Janáček's 'The Cunning Little Vixen', composed 1921–3*

again in the 15th century – were Janáček's own idea
and had the advantage of suggesting the psychology of
Brouček's 'dreams'. But the animal–human links in the
Vixen have less to recommend them and are often
abandoned in performance.

While chary of inventing dialogue – he managed to
dispense with it altogether in the second convict play
performed in *From the House of the Dead* – Janáček
was resourceful in turning indirect into direct speech in
cases of need. Some of the Gamekeeper's musings in
the *Vixen* are in literary Czech (rather than in the
Gamekeeper's homely dialect) because Janáček was
extracting some of the novelist's general reflections. But
he was also capable of daring and imaginative strokes.
In two works, the *Vixen* and *Makropulos*, he added the
deaths of his heroines. In *From the House of the Dead*
he achieved some of his most striking effects by stark
juxtapositions. Luka's tale of his horrific beating is
followed by the return of Petrovič after similar treatment by the prison guards. The torturing and release of
the eagle is juxtaposed respectively and symbolically
with the torturing and release of Petrovič. In all these
later works Janáček's changes alter the slant, and hence
the 'meaning', of the original, and his music completes
the process. He turned Těsnohlídek's lighthearted tale
of a vixen and a gamekeeper into a profound fable that
can comprehend and come to terms with death. The
death of Emilia Marty became one of Janáček's most
magnificent finales, his music investing Čapek's conversation piece with a moving grandeur and monu-
mentality. *From the House of the Dead*, his slackest libretto in terms of events, is fuelled by music of an intense
driving force, startling even for Janáček. His dramatic

instincts, as strong as anything in Puccini, dwarf the few clumsy mistakes (e.g. the implausible timing for Kát'a's suicide). His four last operas are all short, terse to the point of abruptness, yet beautifully paced with a knack of ending an act at exactly the right time.

Jenůfa, alone among the play-based operas, offers several three-dimensional portraits of characters whose spiritual growth through the opera provides its chief subject matter. In *Kát'a* and *Makropulos* Janáček instead concentrated on a single individual, tracing its development against the other characters, which are firmly drawn but static. The intensity and interest of the central portrait in *Makropulos* is such that it overwhelms the complexities of a plot that might otherwise have doomed the opera, one of Janáček's most exciting. It is largely through this narrow focus that *Kát'a*, the *Vixen* and *Makropulos*, though written closely one after the other, are able to evoke such unmistakably different musical worlds.

Janáček inherited his operatic conventions mostly from Smetana and Dvořák and deployed them unquestioningly in his first two operas. Both are number operas, with arias, duets, ensembles and choruses. *The Beginning of a Romance* was originally to have been a Singspiel but its spoken dialogue became upgraded during composition to melodrama and thence to recitative. During the writing of *Jenůfa* Janáček began to show unease about these conventions, perhaps because the prose text and naturalistic character of the play he was adapting did not readily provide for them. There were a few built-in choruses (the recruits' scene, the bridesmaids' song) but the 'duo' Janáček noted at the end of Act 2 of his copy of the play had to be manufactured by combining solo lines, while the substantial concertato for four soloists and chorus in Act 1 grew out of a single line for one character. There was a thoughtful gap of several years between the composition of Acts 1 and 2, after which Janáček's use of 'numbers' became even more sparing. Those he did write were cut down further in his revisions before the vocal score was published – by which time he had become acquainted with some of the most influential models for his middle-period works: Puccini, Strauss and Charpentier's *Louise*. While ensembles persist in *Fate* and *Brouček*, by *Kát'a* and *Makropulos* there are few passages where solo voices combine for more than a bar or two; where they overlap there is usually a 'naturalistic' explanation. Janáček thus became more dependent on the musical monologue and most of his librettos from *Jenůfa* onwards provide many such confessional or narrative monologues. Their frequency in *From the House of the Dead* may be one reason why he was so attracted to this seemingly unoperatic material.

Janáček appeared reluctant to do away with the chorus; even after *Jenůfa* there is plenty of chorus work in *Fate* and *Brouček* (including some thrilling settings of Hussite chorales) but by *Kát'a* he had found a new symbolic role for the chorus, which thereafter is heard, generally off stage, as 'voices of the Volga', 'voices of the forest', etc. In *From the House of the Dead* he tried a new approach. The opera is set in a prison camp and has a body of men on the stage most of the time; Janáček conceived it as a 'chorus opera' in which individual soloists emerge to tell their tales and then retire into the collective identity of the chorus.

From *Jenůfa* onwards, as the set number, the duet

and the ensemble give way to the monologue, the operatic conversation and the symbolic chorus, so Janáček sought other means of formal articulation and the orchestral part became more important, especially in the novel-based operas with their more diffuse plots. In *Brouček*, to some extent, but in particular in the *Vixen* and *From the House of the Dead*, Janáček was able to make tiny scenes cohere by bedding them into an orchestral continuum. He made no use of the leitmotif and only sporadic use of a few reminiscence themes. Instead, his approach was to build up sections – often a whole scene – on a single motif, subjected to variation techniques, sometimes contrasted with other themes and usually forming a type of loose rondo. The second half of Act 2 of the *Vixen* is bonded by the structural arch of the offstage chorus; the first half consists of a set of variations on the theme of the opening prelude. The success of the scheme depended almost entirely on his imaginative treatment of the theme, out of which he was able to coax a wide range of moods and emotions without sacrificing its identity.

With the orchestra taking a more dominant role the voice was released to move freely in a more 'realistic' presentation of the text. A catalyst in this process was Janáček's 'speech-melody' theory. From about 1897 he took down examples of speech in conventional musical notation and studied them, trying to establish the influences of moods and emotions as well as external factors on their rhythm and pitch (some examples which he jotted down during a lecture he heard on Dante are shown in fig.4). He wrote up his 'research' on speech melodies in a variety of ways from stern theoretical articles to lighthearted evocations of tiny scenes brought to life by the inclusion of a snatch of sing-song speech, or an animal's cry, or the din of a tram. Štědroň (*Zur Genesis* ..., 1972) lists ninety-eight articles dealing in some way with speech melody – a practical demonstration of Janáček's belief in their importance to a composer of operas. Speech melodies however were in no sense potential thematic material – an idea which Janáček firmly repudiated – but, rather, study material to help him produce the sung stylizations which came to resemble the irregular patterns of everyday speech. This is not to say that Janáček always set voice parts naturally. The many passages, particularly in *Jenůfa*, where the voice part hugs the accompaniment suggest that some of the voice parts were composed by the

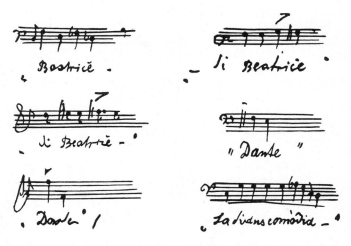

4. *Some examples of speech melodies taken down by Janáček during a lecture on Dante on 6 June 1921*

fitting of words to existing instrumental tunes. An example of this, and how in the course of composition Janáček uncoupled the voice and the orchestra, can be seen in his revision of the voice part of *Jenůfa*. In the original version (1904) of the passage (ex.5) the voice part and the orchestral tune are identical; in the revision (1908) the voice part is detached rhythmically and its two, originally identical, phrases are varied. It is

Ex.5 *Jenůfa*, Act 1

characteristic that the new voice parts begin after the beat. In later works the rhythm becomes even more bunched over the phrase climax, beginning late and ending early, and the voice travels more jaggedly and rapidly over a larger range, with the melodies becoming tighter and more intense.

Janáček had relied on verse librettos in *Šárka* and *The Beginning of a Romance* to provide regular units –

Grandmother, don't be angry

lines of verse – to fit his regular musical phrases. Although he set much store by *Jenůfa*'s being the first opera written to a prose libretto (he was overlooking the claims of Musorgsky and others) the voice and the orchestra had not yet parted company rhythmically and he was obliged to restructure the prose text into metrical units to fit the symmetrical phrases of his music. Rather than repeat this labour himself in his next opera, *Fate*, he instructed his librettist to turn the prose draft into 'Pushkinesque verse'. Only when as a result of the speech-melody theory the rhythmic and melodic detachment of voice and orchestra was regularized did this sort of rhythmic restructuring of the prose become unnecessary, except in passages where an even, regular melody was part of Janáček's characterization plan. (Compare Kát'a's first utterance in the opera, where the text has been metrically modified into regular iambs, to Kabanicha's jagged response, where Janáček set the prose as it stood; ex.1.) Although much of the time voice and orchestra go their own ways thematically, echoes between them provide thematic links and even emotional comments. A further advantage of Janáček's 'detached' voice and orchestra method is the special impact possible at emotionally important moments, for example Marty's great final monologue in *Makropulos*, when voice and orchestra come together in an expansively lyrical, rhythmically united whole.

CHAPTER EIGHT

Writings, reputation

In the Moravian musical life of his time, Janáček was scarcely more important as a composer than as a theorist, teacher and folk music authority. His organ school trained a generation of teachers and administrators, and his work on folksong laid the foundations for a more scholarly approach. His activities in these fields are documented by his considerable literary output. As a writer, his career effectively began in 1883 when he founded the periodical *Hudební listy*, serving as its editor and chief contributor; it was at first (1884–5) a weekly, then (1885–7) a fortnightly, and in its final year a monthly. In it Janáček reviewed the opera performances at the new Provisional Theatre in Brno; most of his reviews are short and deal chiefly with matters of performance, but occasionally – as with Gounod's *Faust* or Rozkošný's *St John's Rapids* – he wrote more detailed introductions, and from time to time he offered observations on the state of opera in Brno. After the demise of *Hudební listy* Janáček published opera reviews in *Moravské listy* (1890–92).

His earliest theoretical writings were also for *Hudební listy*: substantial pieces on topics such as the triad, counterpoint and the concept of tonality, serialized over several issues. Some of the ideas were later developed in his two books on music theory. In spite of his reputation as a composer and his impressive equipment as a theorist, Janáček's theoretical writings had no impact on Czech theoretical thought. His attempt to give his harmony manual wider currency failed when Universal Edition declined to publish it in German. It is neglected partly because it stands outside the tradition of such manuals: its intellectual roots can be found in Janáček's wide-ranging readings in aesthetics and experimental psychology. Further, Janáček's poetic or abstract expression and his home-made terminology often make it difficult to understand.

Janáček was at his most distinctive in his approach to harmony and rhythm. In harmony, the succession of sounds and the momentary confusion when one chord follows another interested him most. The overlapping sounds and the montage of layers in his own works may be seen as reflections of this. Rhythmically he saw a piece as a hierarchy of layers, each with a distinct personality. Janáček also published articles in *Hudební listy* on school music education and on singing teaching, the subject of a further short manual, which treats pitch and rhythm differentiation.

Whereas Janáček's activities as a critic were fitful, confined to two spells between 1884 and 1892, he wrote regularly on folk music for over 40 years, from a review in 1886 to an article published a few months before his death. His most substantial and systematic work came in the introductions he wrote for the collections published with Bartoš and for his posthumous collection of Moravian love-songs. These writings include many of his own folksong notations taken down in almost 50 locations over 25 years.

In 1893 Janáček began writing articles for the new Brno daily, *Lidové noviny*, continuing sporadically over 27 years. From 1921 he contributed regularly, publishing 40 articles in the eight years up to his death. His short *feuilletons* were conceived for a large, popular readership. Some are autobiographical, painting vivid scenes from childhood cut through with reflections from old age; others are pictures of Janáček's environment and tales from his travels; and there are some amusing and wonderfully observed descriptions of animals. Many pieces are springboards for his demonstrations of speech melodies, his most constant theme; others are poetical explanations of the relations between natural and artistic creation. *Lidové noviny* provided the literary sources for three of his major works, *The Diary of One who Disappeared*, *The Cunning Little Vixen* and *Nursery Rhymes*; and it was here too that he wrote some of his rare introductions to his works, *Brouček*, the *Vixen*, the *Glagolitic Mass* and the Sinfonietta. Janáček's *feuilletons* have much the same spirit as his music. His prose comes in short, abrupt phrases, often too compressed and overloaded to reveal its meaning immediately, but with unmistakable energy and force.

In his last decade, Janáček had the satisfaction of seeing himself regarded as Czechoslovakia's leading composer. By his death, *Jenůfa* was established as a repertory opera in Czechoslovakia and in German-speaking Europe, but despite the advocacy of such conductors as Klemperer, Kleiber and Horenstein, none of his later operas achieved more than a few performances outside Czechoslovakia before World War II. It was not until Felsenstein's Berlin production in 1956 that *The Cunning Little Vixen* aroused much interest abroad. No Janáček opera was performed in Britain until 1951, when *Kát'a Kabanová* was presented, but in the 1950s and 1960s Sadler's Wells staged all Janáček's major operas (except *Jenůfa*, the prerogative of Covent

49

Garden); in the late 1970s, with the Decca cycle of recordings and Welsh and Scottish cycles, Janáček's operas achieved wider currency in Britain. They remain less known in the USA.

Until recently Janáček's operas have been heard mostly in reorchestrated versions. Since its 1916 Prague production, *Jenůfa* has always been performed in Karel Kovařovic's orchestration. Similarly, *From the House of the Dead* was known only in a version made after Janáček's death with the orchestration thickened, voice parts displaced and an 'optimistic' ending added. The orchestration of *Kát'a Kabanová* and the *Vixen* was revised by Václav Talich before World War II. His versions persisted in Prague until the 1970s and were perpetuated on the first gramophone recordings. In 1958 Janáček's original ending for *From the House of the Dead* was published as an appendix to the vocal score, and Rafael Kubelik rid the orchestral score of many later accretions. However, he went back only to Janáček's autograph score, ignoring his later additions in the authorized copy. Charles Mackerras recorded the score in its final state (1980). Mackerras also jettisoned the changes to *Jenůfa*, and conducted the original score at the Paris Opéra in 1981, subsequently recording it. The return in the late 1970s and 1980s to Janáček's original orchestration has coincided with the appearance of the first volumes of the collected edition of his works published by Supraphon and Bärenreiter, which has stirred up controversy in Czechoslovakia and elsewhere by its espousal of a notational system that considerably alters the music's appearance. It is evidence, however, of a growing popular and scholarly interest in a composer seen for far too long as an eccentric outside the mainstream.

WORKS

Editions: *L. Janáček: Souborné kritické vydání*, ed. J. Vysloužil and others (Prague, 1978–) [SKV]

All printed works were published in Prague unless otherwise stated; autographs are chiefly in *CS-Bm*, but some are in *A-Wn*; several authorized printer's copies are in Universal Edition archives, Vienna.

Numbers in the right-hand column denote references in the text.

STAGE

(unless otherwise stated, all are operas, first performed at National Theatre, Brno, and published in vocal score)

Title	Acts, Text	Date	First performance	Publication, Remarks	
					23–4, 36–46
Šárka	3, J. Zeyer	1887, rev. 1888, 1918–19, 1924–5	11 Nov 1925	Act 3 orchd O. Chlubna, 1918–19	7, 16, 21, 24, 34, 36, 37, 45
Rákos Rákoczy, folk ballet	1, J. Herben	1891	Prague, National Theatre, 24 July 1891	incl. folk choruses, songs and dances; ed. L. Matějka (Vienna, 1977; Prague, 1978) [reproduced MS]	8
Počátek románu [The beginning of a romance]	1, J. Tichý, after G. Preissová's story	1891	10 Feb 1894	sections later destroyed by Janáček; reconstructed B. Bakala; ed. E. Holis (Prague and Vienna, 1978) [reproduced MS]	8, 9, 34, 36, 37, 40, 45
Její pastorkyňa [Her foster-daughter; Jenůfa]	3, Janáček, after Preissová's play	1894–1903, rev. before 1908	21 Jan 1904	(Brno, 1908); full score rev., reorchd K. Kovařovic (Vienna, 1917); ed. J. M. Dürr (Vienna, 1969)	8–9, 10, 11, 13–14, 15, 16, 17, 21, 22, 23, 24, 25, 26, 33, 34, 35, 36, 37, 40–41, 43–5, 46, 49, 50
Osud [Fate; Destiny]	3, F. Bartošová and Janáček	1903–5; rev. 1906–7	Brno Radio, 18 Sept 1934; Brno, National Theatre, 25 Oct 1958	1958 production = rearr. version by V. Nosek; orig. version ed. V. Nosek (Prague and Vienna, 1978) [reproduced MS]	10, 11, 24, 26, 33, 34, 36, 37, 41, 46
Výlet pana Broučka do měsíce [Mr Brouček's excursion to the moon]	2, Janáček, with addns by F. Gellner, V. Dyk and F. S. Procházka, after S. Čech's novel	1908–17	Prague, National Theatre, 23 April 1920	epilogue act discarded when 2nd excursion added	10–11, 13, 14, 16, 24, 26, 37, 39, 41, 42, 49
Výlet pana Broučka do XV. století [Mr Brouček's excursion to the 15th century]	2, F. S. Procházka, after S. Čech's novel	1917	Prague, National Theatre, 23 April 1920	with 1st excursion as Výlety páně Broučkovy [The excursions of Mr Brouček] (Vienna, 1919)	13
Káťa Kabanová	3, Janáček, after A. N. Ostrovsky, trans. V. Červinka: The Thunderstorm	1919–21	23 Nov 1921	full score (Vienna, 1922), ed. C. Mackerras (Vienna, 1971)	16, 17, 23, 27–8, 33, 34, 35, 37, 40, 41, 46, 49, 50

51

Title	Acts, Text	Date	First performance	Publication, Remarks	
Příhody Lišky Bystroušky [The adventures of the vixen Bystrouška; The cunning little vixen]	3, Janáček, after R. Těsnohlídek's novel	1921–3	6 Nov 1924	(Vienna, 1924)	16, 18, 23, 24, 33, 36, 37, 38, 39, 40, 42, 49, 50
Věc Makropulos [The Makropulos affair]	3, Janáček, after K. Čapek's play	1923–5	18 Dec 1926	(Vienna, 1926); full score, ed. C. Mackerras (in preparation)	16, 23, 29, 35, 36, 37, 39, 40, 41, 46
Z mrtvého domu [From the house of the dead]	3, Janáček, after Dostoyevsky's novel	1927–8	12 April 1930 [version by O. Chlubna, B. Bakala and O. Zitek]	(Vienna, 1926); rev. and reorchd O. Chlubna and B. Bakala, text rev. O. Zítek, vocal score and full score (Vienna, 1930); vocal score with orig. ending as appx (Vienna, 1958)	18, 20, 26, 33, 35, 36, 37, 39, 41, 42, 50
Schluk und Jau, incidental music	G. Hauptmann	1928		sketches of introduction and 3 movts only; 2 movts rev. J. Burghauser, full score (Prague and Vienna, 1978)	18

Projected operas with musical sketches: Gazdina roba [The housewife] (Preissová), 1904, 1907; Paní mincmistrová [The mintmaster's wife] (L. Stroupežnický), 1906–7; Anna Karenina (Tolstoy), 1907 [in Russ.]; Živá mrtvola [The living corpse] (Tolstoy), 1916 — 11

Other projected operas, with only scenario, annotated play or novel etc: Poslední Abencerage (Chateaubriand: Les aventures du dernier des Abencérages), 1885; Andělská sonata [The angel sonata] (J. Merhaut), 1903; Class exercise: Divoška [The Tomboy] (V. Krylov), 1920–21 [corrections to composition pupils' versions]; Dítě [The child] (F. X. Šalda), 1923 — 11

Projected ballets: Pod Radhoštěm [At the foot of Radhošť mountain], 1888–9; Valašské tance [Dances from Valašsko] (V. Kosmák), 1889

SACRED — 4

Fidelis servus, mixed vv, ?1872

Mass, ?1872, lost

Graduale in festo purificationis B.V.M. (Suscepimus), mixed vv, ?1873 rev. 1887, ed. J. Trojan (Prague and Vienna, 1978)

Graduale (Speciosus forma), mixed vv, org, c1874

Introitus (in festo Ss Nominis Jesu), mixed vv, org, c1874

Benedictus, S, mixed vv, org, 1875

Communio, mixed vv, 1875

Exaudi Deus, mixed vv, org, 1875 — 4

Exaudi Deus, mixed vv, 1875; rev. version in Cecilia, iv (1877), suppl. no.3

Regnum mundi, mixed vv, c1878

Sanctus, mixed vv, 1879, lost

Deset českých církevních zpěvů z Lehnerova mešního kancionálu [10 Czech hymns from the Lehner hymmbook for Mass], org with text incipits, 1881 (Brno, 1881–2, 2/1889); arr. TTBB, org, 1881

Ave Maria (Byron, trans. J. Durdík), 1883, Varyto, suppl. to xiii/1 (1890); ed. J. Trojan (Prague, 1979)

Odpočin si [Take your rest] (F. Sušil), male vv, 1894 (1926); SKV: C/1

Hospodine! [Lord have mercy], S, A, T, B, 4vv, 4vv, org, harp, 3 tpt, 4 trbn, tuba, 1896; ed. J. Trojan (Prague and Kassel, 1977)

Slavnostní sbor [Festival chorus] (Sbor k sv. Josefu) [Chorus to St Joseph] (V. Šťastný), TTBB, 1897; ed. J. Trojan (Prague and Vienna, 1978); SKV: C/1

Svatý Václave! [St Wenceslas], org, c1902 [acc. to St Václav hymn]

Constitues, male vv, org, before 1903, rev. 1903, ed. J. Trojan (1971)

Zdrávas Maria (Ave Maria), T, mixed vv, org, 1904; arr. S/T, vn, pf/org, 1904, ed. B. Bakala (Prague and Vienna, 1978)

[7] Církevní zpěvy české vícehlasné z příborského kancionálu [Czech hymns for several vv from the Příbor hymnbook], c1904

Mass, Eb, mixed vv, org, 1907–8, ed. J. Trojan (1972) [Ky, Cr, Ag completed and orchd by V. Petrželka, parts (1946)]

Veni sancte spiritus, TTBB, ?1910; ed. J. Trojan (Prague, 1978)

(unless otherwise stated settings for TTBB of traditional Moravian texts)

Srbská lidová píseň [Serbian folksong], SATB, perf. 27 April 1873, lost 2, 21

Ženich vnucený [The enforced bridegroom], SATB, perf. 27 April 1873, lost [possibly = Srbská lidová píseň]

Oráni [Ploughing], perf. 27 April 1873, no.1 of Ohlas národních písní, pubd in Čtyři lidové mužské sbory (1923); SKV: C/1

Válečná [War song] (anon.), TTBB, tpt, 3 trbn, pf, 1873

Nestálost lásky [The fickleness of love], TTTBB, 1873; ed. J. Trojan (Vienna, 1977; Prague, 1978); SKV: C/1

Osamělá bez těchy [Alone without comfort], 1874, rev. 1898, 1925; ed. J. Trojan (Vienna, 1977; Prague, 1978); SKV: C/1

Divím se milému [I wonder at my beloved], c1875–6, ed. in Ohlas písni národních (Brno, 1937); SKV: C/1

Vínek stonulý [A drowned wreath], c1875–6, ed. in Ohlas písni národních (Brno, 1937); SKV: C/1

Když mne necheš, co je vic? [If you don't want me, what else is there?] (F. L. Čelakovský), perf. 23 Jan 1876 as no.2 of Ohlas národních písni [only T survives]

Láska opravdivá [True love], perf. 23 Jan 1876 as no.3 of Ohlas národních písni, ed. in Ohlas písni národních (Brno, 1937); SKV: C/1

Zpěvná duma [Choral elegy] (Čelakovský), TTBB, pf, before 23 Feb 1876; ed. J. Trojan (Vienna, 1977; Prague, 1978); SKV: C/1

Slavnostní sbor [Festival chorus] (Slavnostní trojsbor) [Festival triple chorus] (K. Kučera), T, T, B, B, SA, TTBB, 1877, ed. J. Trojan (Prague and Kassel, 1972)

Osudu neujdeš [You cannot escape your fate] (S. Kapper), before 29 May 1878; ed. J. Trojan (Vienna, 1977; Prague, 1978); SKV: C/1 7

Na košatej jedli dva holubi seď'á [On the bushy fir tree two pigeons are perched], ?1878–80; SKV: C/1

Píseň v jeseni [Autumn song] (J. Vrchlický), SATB, 1880, ed. B. Štědroň (1951)

Na prievoze [On the ferry], 1883–4; SKV: C/1

Mužské sbory [Male-voice choruses], before 20 June 1885 (Brno, 1886, 2/1924 as Čtveřice mužských sborů [4 male-voice choruses]): 7
Výhrůžka [The warning], Ó lásko [O love], Ach vojna, vojna [Alas the war], Krásné oči tve [Your lovely eyes] (J. Tichý); SKV: C/1

Kačena divoká [The wild duck], SATB, 1885, Zpěvník pro školy střední a měšťanské, ii, ed. B. Žalud (Brno, 1885), 141; ed. B. Štědroň (1951)

Tři mužské sbory [3 male-voice choruses], 1888, ed. M. Venhoda (1959): Loučeni [Parting] (E. Krásnohorská), Holubička [The dove] (Krásnohorská), Žárlivec [The jealous man] (Moravian trad.); SKV: C/1

Naše píseň [Our song] (S. Čech), SATB, orch, 1890, arr. SATB unacc., 1890, ed. B. Štědroň (1951); orch version with new text as Sivý sokol zaletěl [The grey falcon flew away], ?1890 [also used for final septet, no.16, Počátek románu]

Zelené sem seła [I have sown green], SATB, orch, 1892 (arr. pf as Ej, danaj!, 1892]

Což ta naše bříza [Our birch tree] (Krásnohorská), 1893, Památnik Svatopluka (Brno, 1893); repr. in Čtyři lidové mužské sbory (1923); SKV: C/1

Vínek [The wreath], ?1893, pubd in Čtyři lidové mužské sbory (1923)

Už je slúnko z téj hory ven [The sun has risen above that hill], Bar, SATB, pf, perf. 13 May 1894

Čtvero mužských sborů moravských [4 Moravian male-voice choruses], 1904 (1906): Dež viš [If you only knew] (O. Přikryl), Komáři [Mosquitoes], Klekánica [The evening witch] (Přikryl), Rozloučeni [Parting]

Kantor Halfar (P. Bezruč), 1906, rev. 1917 (1923)   12, 21, 25

Maryčka Magdónova (Bezruč), 1906, rev. 1907 (1909)   12, 21, 25

Sedmdesát tisíc [The 70,000] (Bezruč), 1909, rev. 1913 (1923)   12, 21, 25

Perina [The eiderdown], ?1914, pubd in Čtyři lidové mužské sbory (1923) 12

Vlči stopa [The wolf's trail] (Vrchlický), S, SSAA, pf, 1916, ed. J. Ledeč (1968) 12, 13

Hradčanské písničky [Songs of Hradčany] (F. S. Procházka), 1916 (1922), ed. Z. Mrkos (1968): Zlatá ulička [The golden alley], SSAA; Plačíci fontána [The weeping fountain], S, SSAA, fl; Belveder, S, SSAA, harp

Kašpar Rucký (Procházka), S, SSAA, 1916 (1925)   12

Česká legie [The Czech legion] (A. Horák), 1918   21

Potulný šílenec [The wandering madman] (R. Tagore), S, TTBB, 1922 (1925)

Naše vlajka [Our flag] (Procházka), 2 S, TTBB, 1925–6

Sbor při kladeni základního kamene Masarykovy university v Brně [Chorus for laying of foundation stone of Masaryk University in Brno] (A. Trýb), TTBBBB, 1928

CANTATAS

Amarus (Vrchlický), S, T, Bar, SATB, orch, c1897, rev. 1901, 1906, 21–2
vocal score (1938), reproduced MS (1957)   9, 21

53

Otče náš [Our Father] (Moravská Otče náš) [Moravian Our Father], tableaux vivants to paintings by J. Krzesz-Męcina, T, SATB, pf/harmonium, 1901, rev. 1906 for T, SATB, harp, org, ed. B. Štědroň (1963)

Elegie na smrt dcery Olgy [Elegy on the death of my daughter Olga], (M. N. Veveritsa), T, SATB, pf, 1903, rev. 1904, ed. T. Straková (1958) 12, 22

Na Soláni Čarták [Čarták on the Solán] (M. Kurt, pseud. of M. Kunert), T, TTBB, orch, 1911, vocal score (1958), reproduced MS (1958); SKV: B/3 12, 22

Věčné evangelium [The eternal gospel] (Vrchlický), S, T, SATB, orch, 1914, reproduced MS (1958), vocal score (1959) 12, 22

Glagolská mše (Mša slavonija; Glagolskaja missa) [Glagolitic Mass] (church Slav, arr. M. Weingart), S, A, T, B, SATB, orch, org, 1926, vocal score (Vienna, 1928), full score (Vienna, 1930) 17, 22, 24, 33, 34, 49

Když mne nechceš, co je víc? [If you don't want me, what else is there?] (F. L. Čelakovský), song, 1v, pf, ?1869–71

OTHER VOCAL

Jarní píseň [Spring song] (J. Tichý, pseud. of F. Rypáček), song, 1v, pf, 1897, rev. 1905 (Brno, 1905)

Zápisník zmizelého [The diary of one who disappeared] (anon.), song cycle, T, A, SSA, pf, 1917–19 (Brno, 1921) 15, 21, 26, 33, 34, 49

Říkadla [Nursery rhymes] (trad.), 18 pieces and introduction, 2 S, 2 A, 3 T, 2 B, ocarina, fl, fl/pic/a fl, cl, cl/Eb-cl, bn, bn/dbn, db, children's drum, pf, 1927, authorized reduction for 6–9vv, va/vn, pf by E. Stein (Vienna, 1928), full score (Vienna, 1929) [orig. 8 pieces, 3 female vv, cl, pf, 1925; 2 pubd as suppl. to ReM, vii/10 (1926)] 17, 22, 24, 33, 34, 49

Lost: Smrt [Death] (M. N. Lermontov), melodrama, reciter, orch, perf. 27 July 1876; Die Abendschatten (K. Mayer), song cycle, 1v, pf, 1879; song for L. Grill, 1v, pf, 1879; Frühlingslieder (V. Zusner), song cycle, 1v, pf, 1880; songs, 1v, pf, perf. 5 March 1899 5

ORCHESTRAL

Suite, str, 1877 (Brno, 1926) 4
Idylla [Idyll], str, 1878, ed. B. Štědroň (1958) 4
Suite (Serenade), op.3, 1891, ed. B. Štědroň (1958) 8
Adagio, 1891, reproduced MS (1958), ed. O. Chlubna (1964)
Úvod k Její pastorkyni – Žárlivost [Prelude to Jenůfa – Jealousy], ov., 1894, reproduced MS (1958), ed. O. Chlubna (1964), arr. pf 4 hands 1894 [played only as concert ov. during Janáček's lifetime]

Šumařovo dítě [The fiddler's child], ballad (sym. poem) after S. Čech, 1912 (Brno, 1914) 12

Taras Bulba, rhapsody after Gogol, 1915–18, arr. pf 4 hands B. Bakala (1925), full score (1927); SKV: D/7 12, 26

Balada blanická [The ballad of Blaník], sym. poem after Vrchlický, 1920, ed. B. Bakala (1958) 13

Sinfonietta (Vojenská symfonietta; Sletová symfonietta) [Military sinfonietta; Sokol festival sinfonietta], 1926 (Vienna, 1927) 17, 23, 32, 35, 49

Dunaj [The Danube], sym. poem, frag., 1923–8 [completed by O. Chlubna, 1948] 16, 18

Violin Concerto 'Putování dušičky' [Pilgrimage of the soul], sketches, 1927–8 [used in ov. to Z mrtvého domu]

Lost: Scherzo for sym., 1880; Moderato, frag., c1904

See also 'Folk music arrangements'

CHAMBER

Znělka [Fanfare], 4 vn, A, 1875; Znělka, 4 vn, d frag., 1875 [rev. as 4th movt of Suite] 5, 17

Zvuky ku památce Förchtgotta-Tovačovského [Sounds in memory of Arnošt Förchtgott-Tovačovský], 3 vn, va, vc, db, c1875

Romance, vn, pf, 1879, ed. B. Štědroň (1938) [6 others lost]

Dumka, vn, pf, 1880 (1929)

Pohádka [Fairy tale] (Z pohádky) [From a fairy tale], after V. A. Zhukovsky, vc, pf, 1910, rev. 1923 (1924) 11

Presto, vc, pf, ?1910, ed. J. Trojan (1970)

Violin Sonata, 1914–21 (1922), Balada only (Kutná Hora, 1915) 22, 29

String Quartet no.1, after Tolstoy: The Kreutzer Sonata, 1923 (1925, 1975, critical edn. by M. Škampa) [based on pf trio, 1908–9] 11, 16, 17, 22, 23, 26, 29

Mládí [Youth], fl/pic, ob, cl, b cl, hn, bn, 1924 (1925) 16, 22, 23, 24

Pochod Modráčků [March of the Blue Boys], pic, pf, 1924, ed. in Hudební besídka, iv (1927–8), 121, ed. J. Burghauser (1970) [on themes in feuilleton 'Berlin', 1924, and on Mládí; 3rd movt]

Concertino, pf, 2 vn, va, cl, hn, bn, 1926 (1926) 16, 23

Capriccio 'Vzdor' [Defiance], pf left hand, fl/pic, 2 tpt, 3 trbn, t tuba, 1926, ed. J. Burghauser (1953) 17, 23, 24, 35

String Quartet no.2 'Listy důvěrné' [Intimate letters], 1928 (1928, rev. 2/1949, critical edn. by O. Šourek) 18, 20, 22, 23, 29, 30–31, 35

2 short untitled pieces in Janáček's feuilleton 'Berlin', facs., transcr., Fejetony z Lidových novin (Brno, 1958) [orig. in Lidové noviny, 1924]: 1st piece, fl, spinet, 163; 2nd piece, pic, bells, drum, 165

Lost: Sarabanda, str qnt, perf. 1878; 6 romances, vn, pf, 1879; 2 vn sonatas, 1880; Str Qt, 1880; Minuet and Scherzo, cl, pf, perf. 6 Jan 1881; Pf Trio, after Tolstoy: The Kreutzer Sonata, 1908–9 [used in Str Qt no.1]; Komár [The mosquito], vn, pf, frag.   5 11

KEYBOARD
(for pf 2 hands unless otherwise stated)

Prelude, org, g, 1875, ed. M. Buček in *Varhanní skladby* (1976)   4

Varyto (pro plné varhany) [Lyre (for full organ)], org, d, 1875, ed. M. Buček in *Varhanní skladby* (1976)

Chorální fantasie [Chorale fantasia], org, G, 1875, ed. M. Buček in *Varhanní skladby* (1976)

Rondo, frag., 1877

Piece, org, frag., 1878, ed. in V. Helfert: *Leoš Janáček*, i (Brno, 1939), suppl., p.9, no.41

Thema con variazioni (Zdenčiny variace) [Zdenka variations], 'op.1', 1880, ed. V. Kurz (1944); SKV: F/1

2 pieces, org, c1884 (Brno, 1884); ed. J. Reinberger in *Česká varhanní tvorba*, i (1954), 32; ed. J. Trojan (1975)

Na památku [In memoriam], ?1886, ed. J. Trojan (Prague, 1979)

Hudba ke kroužení kuželi [Music for club swinging; Music for gymnastic exercises], 1893 (Brno, 1895; rev. 2/1950 by B. Štědroň); SKV: F/1

Po zarostlém chodníčku [On the overgrown path], 15 pieces [7 orig. for harmonium], 1901–8, i [nos.1–10] (Brno, 1911), ii [nos.11–15] (Brno, 1942) [nos.1, 2 and 10 orig. in *Slovanské melodie*, ed. I. Kolář, v (Ivančice, nr. Brno, 1901); nos.4 and 7 orig. in *Slovanské melodie*, vi (1902); no.11 orig. in *Večery* (30 Sept 1911), suppl. to *Lidové noviny*]; SKV: F/1   11, 23

1. X. 1905 (Z ulice dne 1. října 1905; Sonata) [From the street, 1 Oct 1905], 1905 (1924) [2 movts only, 3rd destroyed]; SKV: F/1   11, 25

V mlhách [In the mists], before 25 Nov 1912 (Brno, 1913; rev. 5/1949 by B. Štědroň); SKV: F/1   11, 23

Vzpomínka [Reminiscence], 1928 (Prague and Brno, 1936), ed J. Trojan (1975) [orig. in *Muzika* (Belgrade, 1928), no.6]; SKV: F/1

Short untitled pieces in Janáček's feuilletons [from *Lidové noviny*], facs., transcr., *Fejetony z Lidových novin* (Brno, 1958): piece in 'Sedm havraní' [7 ravens], 130 [orig. pubd 1922], piece in 'Ústa' [The mouth], 135 [orig. pubd 1923], piece in 'Měl výtečný sluch' [He had an excellent ear], 86 [orig. pubd 1924], piece in 'Toulky' [Rambles], 190 [transcr. only; orig. pubd 1927]

22

Sketches: Skladba o psu Čiperovi [Composition about the dog Čiperal, 1925–8; Čekám Tě [I'm waiting for you], Má milá duše [My dear soul], ?pf, 1928

Lost: Dumka, 1879; Sonata, E♭, 1879; Nocturne, 1879; Smuteční pochod [Funeral march], 1879; 17 fugues, 1879–80; Zdenčin menuet [Zdenka's minuet], 1880; Sonata, 1880; rondos, 1880; Oříšek léskový [The hazelnut], song transcr., 1899; Jarní píseň [Spring song], cycle, 1912

See also 'Folk music arrangements'

FOLK MUSIC ARRANGEMENTS

11

with F. X. Bakeš: Královničky [The little queens], 10 songs, 1v, pf, perf. 21 Feb 1889, ed. B. Štědroň (1954)

with L. Bakešová and X. Běháľková: Národní tance na Moravě [National dances of Moravia], 21 dances, pf 2 hands, 4 hands, some with 1v (Brno, 1891–3), ed. B. Štědroň (1953)

Ej, danaj!, pf, 1892; arr. SATB, orch as Zelené sem seľa [I have sown green], 1892; SKV: F/1

Ukvalská lidová poesie v písních [Hukvaldy folk poetry in songs], 13 songs, 1v, pf, 1898 (Brno, 1899)

Ukvalské písně [Hukvaldy songs], 6 songs, 4vv, 1899, ed. B. Štědroň (1949)

[2] Moravské tance [Moravian dances], pf, c1904 (Brno, 1905); SKV: F/1

26 Balad lidových [26 folk ballads]: 6 národních písní jež zpívala Gabel Eva [6 folksongs which Eva Gabel sang], 1v, pf, perf. 1909 (1922); [7] Lidová nokturna [Folk nocturnes], 2 female vv, pf, before May 1906 (1922); [8] Písně detvanské (Zbojnické balady) [Detvan brigand songs], 1v, pf, 1916 (1950); 5 národních písní [5 folksongs], T, TTBB, pf/harmonium, 1916–17 (1950)

Podme, milá podme [Come, dearest], 1v, pf, 1911, facs. in L. Janáček and P. Váša: 'Z nové sbírky národních písní moravských' [From the new collection of Moravian folksongs], *Večery* (23 Dec 1911) [*Lidové noviny*, suppl. no.19]; repr. in L. Janáček: *Fejetony z Lidových novin* (Brno, 1958), 299

5 Moravian dance songs: Ten ukvalský kostelíček [This Hukvaldy chapel]; Tovačov, Tovačov; Pilařská [Saw dance]; Aj, ženy [Oh you women]; Krajcpolka, 1v, pf, 1908–12, ed. O. and F. Hrabalovi (1979) [1–4, ed. J. Ceremuga (Prague and Kassel, 1978); 5, facs. in *Večery* (17 Feb 1912) [*Lidové noviny*, suppl. no.20]; repr. in L. Janáček: *Fejetony z Lidových novin* (Brno, 1958), 314]

4 folk ballads: Rodinu mám [I have a family], Seděl jeden vězeň [A

prisoner sat in jail], Rychtarova Kačenka [The mayor's Kačenka], Tam dole na dole [Down there in the mine], 1v, pf, 1908–12, ed. J. Trojan (1980)

Slezské písně (ze sbírky Heleny Salichové) [Silesian songs from Helena Salichová's collection], 10 songs, 1v, pf, 1918 (Brno, 1920)

Moravské lidové písně [Moravian folksongs], 15 songs, pf with text, before 1 Jan 1922, ed. B. Štědroň (1950) [?no.1 orig. Ej], duby, duby [Oh the oaks], folksong, 1v, pf, ed. J. Ceremuga (Prague and Kassel, 1978)]

Radujte se všichni [Rejoice all of you], Sklenovské pomezí [Sklenov border country], Pod'te, pod'te děvčátka [Come girls], pf with text, 1898–9, rev. 1923, ed. in L. Janáček: 'Starosta Smolík' [Mayor Smolík], Lidové noviny (18 March 1923]; repr. in L. Janáček: Fejetony z Lidových novin (Brno, 1958), 23

Numerous folkdances, orch, some with mixed vv, 1888–92, used in different versions: in Valašské tance [Dances from Valašsko], 2 pubd as op.2, (1890); Hanácké tance [Dances from Haná]; Moravské tance [Moravian dances], (1958), ed. J. Burghauser (Prague and Mainz, 1971); Rákos Rákoczy, ballet, perf. 1891; Počátek románu [The beginning of a romance], opera, 1891; Suite, op.3 (1958); Lašské tance [Dances from Lašsko], 1893 (1928), SKV: D/4; Kozáček [Cossack dance], 1899, reproduced MS (1958), ed. J. Burghauser (Prague and Kassel, 1977); Kolo Srbské [Serbian reel], perf. 1900, reproduced MS (1958), ed. J. Burghauser (Prague and Kassel, 1977); many arr. pf in Národní tance na Moravě [National dances of Moravia]

Numerous folkdances, pf, 2 hands, 4 hands, some lost, some incorporated in Národní tance na Moravě [National dances of Moravia], 1888–95: in 20 Lašských tanců z Kozlovic [Lašsko dances [8, 34] from Kozlovice], Lašské tance z Kunčic pod Ondřejníkem [Lašsko dances from Kunčice beneath Ondřejník], Hanácké tance [Dances from Haná], Valašské tance z Jasenice [Valašsko dances from Jasenice]

FOLKSONG EDITIONS

with F. Bartoš: Kytice z národních písní moravských [A bouquet of [8, 48] Moravian folksongs], 1v: 174 songs, unacc. (Telč, 1890, rev. 3/1901 with 21 addl Czech and Slovak songs, 4/1953, ed. A. Gregor and B. Štědroň); 53 songs, pf acc. (Telč, 1892–1901, 2/1908 as Moravská lidová poesie v písních [Moravian folk poetry in songs], 4/1947, ed. B. Štědroň)

with F. Bartoš: Národní písně moravské v nově nasbírané [Moravian [8, 48] folksongs newly collected], 1899 (1899–1901)

with P. Váša: Moravské písně milostné [Moravian love-songs], 1928 (1930–36)

OTHER ARRANGEMENTS

A. Dvořák: Moravské dvojzpěvy [Moravian duets], mixed vv, pf, 4 perf. 1877, 2 more perf. 1884 (1939) [printed privately]; (Prague, 1978)
F. Liszt: Messe pour orgue, arr. SATB, org, 1901; ed. J. Burghauser (Prague and Vienna, 1978)
E. Grieg: Olav Trygvason, arr. solo vv, mixed vv, orch, c1902
trad.: Narodil se Kristus Pán [Born is Christ the Lord], pf acc., after carol melody in Lehner hymnbook, repr. in L. Janáček: 'Světla jitřní', Lidové noviny (24 Dec 1909); repr. in L. Janáček: Fejetony z Lidových novin (Brno, 1958), 19
J. A. Komenský: Ukolébavka [Lullaby], pf acc. added, ed. F. Pražák in Kniha Komenského (Brno, 1920)
J. Haydn: Austrian national anthem (Rakouská hymna), 1v, org

WRITINGS
[47–8]

THEORY, TEACHING

O skladbě souzvukův a jejich spojův [The composition of chords and their connections] (Prague, 1897); ed. in Hudebně teoretické dílo, i (1968), 183–296
Návod pro vyučování zpěvu [Singing teaching manual] (Brno, 1899); SKV: H/2
Úplná nauka o harmonii [Complete harmony manual] (Brno, 1911– [48] 12, 2/1920); ed. in Hudebně teoretické dílo, ii (1974), 169–328
ed. M. Hanák: 'Z přednášek Leoše Janáčka o sčasování a skladbě' [From Janáček's lectures on rhythm and composition], Leoš Janá-

[47–50]

ček: sbornik stati a studii (Prague, 1959), 137–74; see HRo, i (1924–5) and Tempo, xx (Prague, 1947–8), 235 for further extracts from Janáček's lectures
Hudebně teoretické dílo [Music theory works], Janáčkův archiv, ii/2, ed. Z. Blažek: i Spisy, studie a dokumenty [Writings, studies and documents] (Prague, 1968); ii Studie. Úplná nauka o harmonii [Studies, Complete harmony manual] (Prague, 1974) [incl. most of Janáček's writings on music theory; see also Fejetony z Lidových novin (1958), 197 for 'Nový proud v theorii hudební' (A new current in musical theory)]

'Objektivní hodnota hudebního díla' ['The objective value of a musical work], OM, vi (1974), 216 [repr. Janáček's notes for a lecture, c1915]

Z. Blažek, ed.: 'Leoš Janáček: teorie nauk o harmonii' [Theory of the teaching of harmony], OM, xi (1979), 289 [Janáček's lectures to the composition master class at the Brno Conservatory 1919–20]
other writings on theory and teaching listed in ČSHS

FOLK MUSIC   48

O lidové písni a lidové hudbě [Folksong and folk music], Janáčkův archiv, ii/1, ed. J. Vysloužil (Prague, 1955) [complete list of Janáček's writing on folk music on p.81; most repr. here]

SPEECH MELODY   42–3, 43

List of 98 items in which Janáček discussed speech melody given in
B. Štědroň: Zur Genesis von Leoš Janáčeks Oper Jenůfa (Brno, 2/1972), 149; about half repr. in Fejetony z Lidových novin (1958) and O lidové písni (1955)

CRITICISM AND ANALYSIS   48

L. Firkušný: Leoš Janáček kritikem brněnské opery [Janáček as a critic of the Brno Opera] (Brno, 1935) [reprs. of Janáček's opera reviews in Hudební listy, 1884–88]

B. Štědroň: 'Leoš Janáček kritikem brněnské opery v letech 1890–1892' [Janáček as a critic of the Brno Opera 1890–92], Otázky divadla a filmu, i (Brno, 1970), 207–48 [with Eng. summary]; Ger. trans. in Leoš Janáček-Gesellschaft: Mitteilungsblatt 1971, nos.3–4; 1972, nos.1–2 [lists all Janáček's articles in Moravské listy (1890–92) and reprs. those on opera; 2 others repr. in HRo, vii (1954), 640]

Reviews of Fibich works repr. in Zdeněk Fibich: sborník dokumentů a studií, ed. A. Rektorys, ii (Prague, 1952), 304ff; analyses of Dvořák's tone poems repr. in Musikologie, v (1958), 324–59; introduction to Tchaikovsky's Pikovaya dama (1896) repr. in Fejetony z Lidových novin (1958), p.210; facs. of MS analysis of Debussy's La mer (1921) repr. as suppl. to M. Štědroň: 'Janáček, verismus a impresionismus', Časopis Moravského musea: vědy společenské, liii–liv (1968–9), 125
other reviews listed in ČSHS; summary of Janáček's newspaper and periodical articles on other composers in Leoš Janáček et musica europaea: Brno III 1968, 151

49

AUTOBIOGRAPHY, MEMOIRS, INTERVIEWS

A. Veselý, ed.: Leoš Janáček: Pohled do života i díla [A view of the life and works] (Prague, 1924) [autobiography]

J. Racek and L. Firkušný: Janáčkovy feuilletony z L.N. [Janáček's feuilletons from Lidové noviny] (Brno, 1938; rev. and enlarged 2/1958 as L. Janáček: Fejetony z Lidových novin; Ger. trans., 1959) [incl. facs. of several short compositions, 1958 edn. adds transcrs.; 1958 edn. adds 7 feuilletons and omits 1]

J. Racek, ed.: Leoš Janáček: Triptychon (Prague, 1948) [3 feuilletons]

V. and M. Tausky, ed. and trans.: Leoš Janáček: Leaves from his Life (London, 1982) [30 feuilletons in Eng.]

Speeches, interviews etc: many listed with orig. pubn details in T. Strakové, ed.: Leoš Janáček: Musik des Lebens (Leipzig, 1979), 217ff; modern repr. in Sborník prací filosofické fakulty brněnské university, H4 (1969), 121; OM, ii (1970), 105; vi (1974), 224; Leoš Janáček: Mikromeditäiön, ed. A. Buchner (Prague, 1974); Ger. trans. or orig. in Leoš Janáček-Gesellschaft: Mitteilungsblatt 1979, no.3, 1980, nos.2–3; T. Strakové, ed.: Leoš Janáček: Musik des Lebens (Leipzig, 1979)

Janáček's introductions to his own compositions: Dalibor, xxxix (1906–7), 49 [Jenůfa]; HR, vii (1913–14), 203 [The Fiddler's Child]; HR, ix (1916), 245 [Jenůfa]; HR, xiii (1920), 177 [Brouček]; Pult und Taktstock, ix (1927), 63 [Concertino]; Lidové noviny (23 Dec 1917) [Brouček], (1 Nov 1924) [Cunning Little Vixen], (27 Nov 1927) [Glagolitic Mass], (24 Dec 1927) [Sinfonietta], all repr. in Fejetony z Lidových novin (1958); HRo, vii (1954), 639 [On the Overgrown Path]; OM, iii (1971), 106 [Jenůfa]; OM, vi (1974), 199 [Amarus], 207 [general survey, 1928]

49

COLLECTIONS

B. Štědroň, ed.: Janáček ve vzpomínkách a dopisech [Janáček in reminiscences and letters] (Prague, 1946; rev. Eng. trans., 1955 as Leoš Janáček: Letters and Reminiscences; rev. Ger. trans., 1955)

T. Strakové, ed.: Leoš Janáček: Musik des Lebens: Skizzen, Feuilletons, Studien (Leipzig, 1979)

J. Šeda: Leoš Janáček: Hledal jsem proudkem prameniou vodu [I searched with a rod for the source of the water] (Prague, 1982)

CORRESPONDENCE

Janáčkův archiv, ed. J. Racek, 1st ser.: i Korespondence Leoše Janáčka s Artušem Rektorysem (Prague, 1934, enlarged 2/1949 = iv); ii

Korespondence Leoše Janáčka s Otakarem Ostrčilem (Prague, 1948); iii Korespondence Janáčka s F. S. Procházkou (Prague, 1949); v Korespondence Leoše Janáčka s libretisty Výletů Broučkových [Janáček's correspondence with the Brouček librettists] (Prague, 1950); vi Korespondence Leoše Janáčka s Gabrielou Horvátovou (Prague, 1950); vii Korespondence Leoše Janáčka s Karlem Kovařovicem a ředitelstvím Národního divadla [Janáček's correspondence with Kovařovic and the directors of the National Theatre] (Prague, 1950); viii Korespondence Leoše Janáčka s Marií Calmou a MUDr. Františkem Veselým (Prague, 1951); ix Korespondence Leoše Janáčka s Maxem Brodem (Prague, 1953) [i–vii, ed. A. Rektorys; viii–ix, ed. A. Rektorys and J. Racek]

J. Racek: Bratři Mrštíkové a jejich citový vztah k Leoši Janáčkovi a Vítězslavu Novákovi [The Mrštík brothers and their emotional ties with Janáček and Novák] (Brno, 1940) [incl. correspondence]

B. Štědroň: 'Janáčkovy "Listy důverné"' [Janáček's 'Intimate Letters'], HRo, vi (1953), 608 [to Stösslová; see also J. Slavický, ed.: Listy důverné (Prague, 1966)]

T. Straková, ed.: František Bartoš a Leoš Janáček: vzájemná korespondence (Gottwaldov, 1957)

I. Stolařík, ed.: Jan Löwenbach a Leoš Janáček: vzájemná korespondence (Opava, 1958)

B. Štědroň: 'Ke korespondenci a vztahu Leoše Janáčka a Karla Kovařovice' [Correspondence and relations between Janáček and Kovařovic], Sborník prací filosofické fakulty brněnské university, F6 (1960), 31–69

Z. E. Fischmann: 'Janáček's London Visit', Czechoslovakia Past and Present, ii, ed. M. Rechcigl, jr (The Hague, 1968), 1336 [incl. extracts from Janáček–Rosa Newmarch correspondence]

Leoš Janáček: dopisy Zdence [Janáček: letters to Zdenka], ed. F. Hrabal (Prague, 1968) [trans. from Ger. by O. Fiala]

B. Štědroň: 'Janáčkova korespondence s Universal-Edition v letech 1916–18 tykající se Její pastorkyně' [Janáček's correspondence with Universal Edition 1916–18 concerning Jenůfa], Otázky divadla a filmu, ii (Brno, 1971), 249–312; Ger. trans. in Leoš Janáček-Gesellschaft: Mitteilungsblatt 1976, 1977, 1978; see also Sborník prací filozofické fakulty brněské univerzity, H17 (1982), 41 [to Universal Edition over proposed publication of harmony manual]

J. Snížková: 'Janáčkovy dopisy na Bertramce' [Janáček's letters at the Bertramka], Bertramka: Věstník Mozartovy obce v ČSSR, xix (1978), 2

Further Janáček correspondence in Musikologie, i (1938), 130 [with Brno National Theatre]; ibid, ii (1948), 171 [with H. Vojáček]; ibid, iii (1955), 465 and v (1958), 171 [with Hostinský]; ibid, iv (1955), 246 [on folk music activities]; in Smetana, xxxv (1941), 58, 72, 89 [with F. Neumann]; in Tempo, xx (Prague, 1947–8), 239 [with F. Skácelík; with the Olomouc Žerotín choral society]; in Sborník prací filosofické fakulty brněnské university, F3 (1959), 80 [with A. Veselý]; in Slezský sborník, lxi (1963), 103 [with I. Wurm]; in OM, iii (1971), 112 [over The 70,000 and Glagolitic Mass]; in OM, vi (1974), 32 [with J. Kunc]; in Mitteilungen des Österreichischen Staatsarchivs (1972), no.25, p.409 [with H. Gregor]; in Časopis Slezského muzea, xxiii (1974), ser. B, 72 [with J. N. Polášek]; in Sborník prací pedagogické fakulty v Ostravě, D14 (1978), 103 [with J. Vluka and J. Vogel]; in ÓMz, xxxiv (1979), 285 [with Mahler]; in Leoš Janáček-Gesellschaft: Mitteilungsblatt 1981, no.2 [with Klemperer]; in J. Tyrrell: Leoš Janáček: Kát'a Kabanová (Cambridge, 1982), 62, 91, 136

BIBLIOGRAPHY

CATALOGUES, BIBLIOGRAPHIES, DISCOGRAPHIES, LISTS OF PERFORMANCES

J. Racek, ed.: *Leoš Janáček: obraz života a díla* [A picture of Janáček's life and works] (Brno, 1948) [incl. list of works (T. Straková and V. Veselý), 31–61; list of writings (T. Straková), 55; systematic bibliography (O. Fric), 62; iconography (J. Raab), 89]

Leoš Janáček na světových jevištích [Janáček on the world stages] (Brno, 1958) [exhibition catalogue; incl. list, by town, of performances of Janáček operas outside Czechoslovakia, 15]

V. Telec: *Leoš Janáček 1854–1928: výběrová bibliografie* [Select bibliography] (Brno, 1958)

B. Štědroň: *Dílo Leoše Janáčka: abecední seznam Janáčkových skladeb a úprav* [Janáček's works: an alphabetical catalogue of Janáček's compositions and arrangements] (Prague, 1959; Eng. trans., 1959, as *The Work of Leoš Janáček*; Russ. trans., 1959; Ger. trans., *BMw*, ii, 1960, pp.120–53, iii, 1961, pp.34–77)

ČSHS [list of works and bibliography to 1962]

W. D. Curtis: *Leoš Janáček* (Utica, NY, 1978) [discography]

J. Kratochvílová: *Dílo Leoše Janáčka: výběrová bibliografie* [select bibliography] (Brno, 1978)

Z. Tomanová: 'Leoš Janáček a Národní divadlo v předhledech' [Janáček and the National Theatre surveyed], *HRo*, xxxi (1978), 234 [detailed list of productions at Prague National Theatre]

J. Procházka: *Hudební dílo Leoše Janáčka* [Janáček's musical works] (Frýdek-Místek, 1979) [chronological list of works, newly numbered; chronological list of editions 1877–1930]

S. Přibáňová: 'Productions of "Kát'a Kabanová" ', in J. Tyrrell: *Leoš Janáček: Kát'a Kabanová* (Cambridge, 1982), 209

ICONOGRAPHY

J. Raab: 'Janáčkova ikonografie', *Leoš Janáček: obraz života a díla*, ed. J. Racek (Brno, 1948), 89 [documented list]

B. Štědroň: *Leoš Janáček v obrazech* [Janáček in pictures] (Prague, 1958, enlarged 2/1980)

J. Sudek: *Janáček – Hukvaldy* (Prague, 1971) [Hukvaldy countryside and Janáček memorials]

T. Straková, ed.: *Iconographia janáčkiána* (Brno, 1975)

P. Eckstein, ed.: *Leoš Janáček a Národní divadlo* [Janáček and the National Theatre] (Prague, 1978)

S. Jareš: 'Obrazová dokumentace Janáčkovy Její pastorkyně v Národním divadle roku 1916' [Pictorial documentation of *Jenůfa*, National Theatre, 1916], *HV*, xv (1978), 358

EDITORIAL PROBLEMS

'Protokol pracovní porady o problémech edic díla Leoše Janáčka' [Minutes of the working committee on problems of the editions of Janáček's works], *Sborník prací filosofické fakulty brněnské university*, H7 (1972), 97–154

J. Burghauser: 'K problematice Janáčkova zápisu tónových výšek' [On the problems of Janáček's pitch notation], *OM*, x (1978), 149

Prefaces and critical commentaries to all volumes of *Leoš Janáček: souborné kritické vydání* (Prague and Kassel, 1978–)

J. Burghauser and K. Šolc: *Leoš Janáček: souborné kritické vydání: ediční zásady a směrnice* [Janáček complete critical edition: editorial principles and guidelines] (Prague, 1979; Ger. summary, 1979)

Discussion of vol.i of complete critical edn., *HRo*, xxxiii (1980), 34, 180, 233, 285, 334, 375; xxxiv (1981), 422; xxxv (1982), 136; *HV*, xvii (1980), 171; *Notes*, xxxvii (1980–81), 942

SPECIAL PERIODICAL ISSUES, COLLECTIONS OF ESSAYS,
CONFERENCE REPORTS

'Janáčkův sborník' [Janáček volume], *HRo*, i/3–4 (1924–5)

'K jubileu Leoše Janáčka' [Janáček jubilee], *Listy Hudební matice*, iv/1–2 (1924–5)

'Janáčkovo číslo' [Janáček number], *HRo*, iv/4–8 (1928)

'Věnováno Janáčkovi' [Dedicated to Janáček], *Klíč*, iii (1932–3), 241

J. Racek: *Leoš Janáček: poznámky k tvůrčí mu profilu* [Observations towards a profile of Janáček as a creative artist] (Olomouc, 1936) [essays]

'Leoš Janáček, zřený svými současníky' [Janáček seen by his contemporaries], *Rytmus*, viii/7 (1942)

Tempo, xx (Prague, 1947–8), 235

J. Racek, ed.: *Leoš Janáček: obraz života a díla* [A picture of Janáček's life and work] (Brno, 1948) [essays, worklist, bibliography, iconography]

V. Helfert: *O Janáčkovi* [About Janáček], ed. B. Štědroň (Prague, 1949) [collected essays and articles]

HRo, vii/15 (1954)

J. Racek and others, eds.: 'K stému výročí narození Leoše Janáčka 1854–1954' [The 100th anniversary of the birth of Leoš Janáček 1854–1954], *Musikologie*, iii (1955)

I. Stolařík, ed.: *Leoš Janáček: Ostravsko k 30. výročí úmrtí: sborník vzpomínek* [Ostravsko on the 30th anniversary of Janáček's death: collection of reminiscences] (Ostrava, 1958)

Leoš Janáček a soudobá hudba: mezinárodní hudebně vědecký kongres, Brno 1958 [Janáček and contemporary music: international musicological conference, Brno 1958] (Prague, 1963)

Bibliography

V. Nosek, ed.: *Opery Leoše Janáčka na brněnské scéné* [Janáček's operas on the Brno stage] (Brno, 1958) [production pictures and an essay on each opera, incl. O. Chlubna on *Šárka* and his revision of *From the House of the Dead*] *HRo*, xi/18 (1958)

Leoš Janáček: sborník statí a studií (Prague, 1959) [collection of articles and studies]

Sborník: Janáčkovy akademie múzických umění, v (1965) [articles on performance and interpretation]

Operní dílo Leoše Janáčka: sborník příspěvků z mezinárodního symposia Brno, říjen 1965 [Janáček's operatic works: collection of contributions to the international symposium at Brno, October 1965], Acta janáčkiana, i, ed. T. Straková and others (Brno, 1968)

R. Pečman, ed.: *Colloquium Leoš Janáček et musica europaea: Brno 1968* (Brno, 1970) [papers for cancelled 3rd Brno international conference]

Leoš Janáček-Gesellschaft: Mitteilungsblatt (Zurich, 1969–)

'Leoš Janáček: osobnost a dílo: sborník studií a dokumentů' [Personality and works: collection of studies and documents], *OM*, vi/5–6 (1974)

'Leoš Janáček ve škole' [Janáček as teacher; report of Hukvaldy conference, 1975], *OM*, viii/7–8 (1976)

HV, xv/4 (1978)

OM, x/5–6 (1978)

'Katedra hudební výchovy pedagogické fakulty v Ostravě k 50. výročí smrti Leoše Janáčka dne 12. srpna 1928 v Ostravě' [The music education department of the education faculty in Ostrava on the 50th anniversary of Janáček's death on 12 August 1928 in Ostrava], *Sborník prací pedagogické fakulty v Ostravě*, D14 (1978), 81–109 [studies, correspondence, newspaper articles]

R. Pečman, ed.: *Colloquium Leoš Janáček ac tempora nostra: Brno 1978* (Brno, 1983) [papers for 13th Brno international conference]

H.-K. Metzger and R. Riehn, eds.: *Leoš Janáček*, Musik-Konzepte, no.7 (Munich, 1979) [articles, worklist, bibliography]

K. Steinmetz, ed.: *Janáčkiana '78 a '79: Sborník materiálů z hudebněvědeckých konferencí konaných v Ostravě v rámci Janáčkova máje 31. května – 2. června 1978 a 28.–30. května 1979* [Papers of the musicological conferences held in Ostrava 31 May–2 June 1978 and 28–30 May 1979] (Ostrava, 1980)

Z. Zouhar, ed.: *Živý Janáček: sborník z hudebně teoretické konference JAMU* [Living Janáček: papers of the music theory conference held at the Janáček Academy of Music, Brno] (Brno, 1980)

J. Knaus, ed.: *Leoš Janáček – Materialien: Aufsätze zu Leben und Werk* (Zurich, 1982)

MEMOIRS

HRo, iv/4–8 (1928) [incl. recollections by Janáček's pupils]

O. Chlubna: 'Vzpomínky na Leoše Janáčka' [Memories of Janáček], *Divadelní list Zemského divadla v Brně*, vii (1931–2), 101, 125, 169, 172, 289

J. M. Květ, ed.: *Živá slova Josefa Suka* [in Suk's own words] (Prague, 1946), 47ff

B. Štědroň, ed.: *Janáček ve vzpomínkách a dopisech* [Janáček in reminiscences and letters] (Prague, 1946; rev. Eng. trans., 1955 as *Leoš Janáček: Letters and Reminiscences*; rev. Ger. trans., 1955)

J. B. Foerster: 'Leoš Janáček ve Vídni' [Janáček in Vienna], J. B. Foerster: *Poutník v cizině* (Prague, 1947), 265

R. Smetana: *Vyprávění o Leoši Janáčkovi* [Stories about Janáček] (Olomouc, 1948)

S. Tauber: *Můj hudební svět . . . vzpomínky* [My musical world . . . reminiscences] (Brno, 1949), 67, 101

I. Stolařík, ed.: *Leoš Janáček: Ostravsko k 30. výročí úmrtí* [The Ostrava district on the 30th anniversary of Janáček's death] (Ostrava, 1958)

M. Trkanová: *U Janáčků: podle vyprávění Marie Stejskalové* [At the Janáčeks: after the account of Marie Stejskalová (Janáček's housekeeper)] (Prague, 1959, 2/1964)

V. Lébl: *Vítězslav Novák: život a dílo* (Prague, 1964), 370 [extracts from unpubd vol.ii of Novák's memoirs]

M. Brod: 'Erinnerungen an Janáček', *Beiträge 1967*, ed. K. Roschitz (Kassel, 1967), 30

F. Kožík *Po zarostlém chodníčku* [On the overgrown path] (Prague, 1967) [reminiscences of Křička, Horvátová, Kundera, Kunc, Mikota]

J. Racek: *Leoš Janáček v mých vzpomínkách* [Janáček as I remember him] (Prague, 1975) [chiefly contemporary reviews]

R. Kvasnica: 'Leoš Janáček jak učitel ve vzpomínkách Rudolfa Kvasnici' [Janáček as teacher in the reminiscences of Rudolf Kvasnica], *OM*, vi (1974), 172; Ger. trans. in *Leoš Janáček – Materialien*, ed. J. Knaus (Zurich, 1982), 65

G. Pivoňka: 'Janáček očima mých vzpomínek' [Janáček in my reminiscences], *OM*, viii (1976), 234

J. V. Sládek: *Hukvaldské miniatury* (Ostrava, 1979) [incl. Sládek's and his father's reminiscences of Janáček]

F. M. Hradil: *Hudebníci a pěvci v kraji Leoše Janáčka* [Musicians and singers in Janáček country] (Ostrava, 1981) [incl. Hradil's reminiscences of Janáček]

LIFE AND WORKS

M. Brod: *Leoš Janáček: život a dílo* [Life and works] (Prague, 1924; Ger. orig., 1925, rev. 2/1956)

Bibliography

D. Muller: *Leoš Janáček* (Paris, 1930/*R*1975)

A. E. Vašek: *Po stopách dra Leoše Janáčka* [On the track of Dr Leoš Janáček] (Brno, 1930)

V. Helfert; *Leoš Janáček*, i (Brno, 1939)

J. Vogel: *Leoš Janáček: Leben und Werk* (Prague, 1958; Eng. trans., 1962, rev. 2/1981; Cz. orig., 1963) [expanded version of *Leoš Janáček: dramatik* (Prague, 1948)]

J. Šeda: *Leoš Janáček* (Prague, 1961)

J. Racek: *Leoš Janáček: Mensch und Künstler* (Leipzig, 1962, 2/1971; Cz. orig., 1963, as *Leoš Janáček: člověk a umělec*)

H. Hollander: *Leoš Janáček: his Life and Work* (London, 1963: Ger. orig., 1964)

M. Černohorská: *Leoš Janáček* (Prague, 1966) [in Eng.; also in Fr., Ger., Russ., 1966]

B. Štědroň: *Leoš Janáček: k jeho lidskému a uměleckému profilu* [Janáček's image as man and artist] (Prague, 1976) [incl. chronology of life and works in Cz. and Ger. and bibliography, chiefly of post-1965 literature]

G. Erismann: *Janáček ou La passion de la vérité* (Paris, 1980)

I. Horsbrugh: *Leoš Janáček: the Field that Prospered* (Newton Abbot, 1981)

J. Vysloužil: *Leoš Janáček: für Sie porträtiert* (Leipzig, 1981)

K. Honolka: *Leoš Janáček: sein Leben, sein Werk, seine Zeit* (Stuttgart and Zurich, 1982)

S. Přibáňová: *Leoš Janáček* (Prague, 1984)

BIOGRAPHICAL AND HISTORICAL STUDIES

L. Kundera: *Janáček a Klub přátel umění* [Janáček and the Friends of Art Club] (Olomouc, 1948)

——: *Janáčkova varhanická škola* [Janáček's organ school] (Olomouc, 1948)

J. Procházka: *Lašské kořeny života i díla Leoše Janáčka* [The Lašsko roots in Janáček's life and works] (Prague, 1948)

J. Racek: 'Janáček a Praha' [Janáček and Prague], *Musikologie*, iii (1955), 11–50

P. Novák: 'Poslední vůle Leoše Janáčka' [Janáček's last will], *Časopis Matice moravské*, lxxv (1956), nos.3–4, p.279

——: 'Odkaz Leoše Janáčka filosofické fakultě' [Janáček's bequest to the arts faculty], *Sborník prací filosofické fakulty brněnské university*, F1 (1957), 77–101

T. Straková: 'Setkání Leoše Janáčka s Gabrielou Preissovou' [Janáček's encounter with Gabriela Preissová], *Časopis Moravského musea: vědy společenské*, xliii (1958), 145

P. Vrba: 'Janáčkova první cesta do Ruska roku 1896' [Janáček's first journey to Russia in 1896], *Slezský sborník*, lvii (1959), 464

Janáček

———: 'Ruský kroužek v Brně a Leoš Janáček' [The Russian circle in Brno and Janáček], *Slezský sborník*, lviii (1960), 71

———: 'Janáčkova ruská knihovna' [Janáček's Russian library], *Slezský sborník*, lx (1962), 242

J. Fukač: 'Leoš Janáček a Zdeněk Nejedlý', *Sborník prací filosofické fakulty brněnské university*, F7 (1963), 5

J. Racek: 'Janáčeks Studienaufenthalt in Leipzig in den Jahren 1879–1880', *1843–1968: Hochschule für Musik Leipzig* (Leipzig, 1968), 187; enlarged Cz. version in *Časopis Moravského musea: vědy společenské*, lxii (1977), 75

O. Chlubna: 'Vzpomínky na Janáčkův případ' [Recollections about Janáček's case], *OM*, ii (1970), 281; Ger. trans. in *Leoš Janáček – Materialien*, ed. J. Knaus (Zurich, 1982), 68 [on the demise of Janáček's organ school]

B. Štědroň: 'Leoš Janáček a Rustý kroužek v Brně' [Janáček and the Russian circle in Brno], *Program* [*Státního divadla v Brně*], xliv (1972–3); xlv (1973–4)

V. Horák: 'Janáček a Pivoda', *OM*, viii (1976), 199

B. Štědroň: 'Janáček a Hukvaldy', *OM*, viii (1976), 195

V. Gregor: 'Leoš Janáček a Cyril Metoděj Hrazdira', *Sborník prací pedagogické fakulty v Ostravě*, D14 (1978), 83

J. Procházka: 'Leoš Janáček, Václav Talich a Česká filharmonie', *OM*, x (1978), no.8, p.184, p.xiii

O. Pulkert: 'Dramatické dílo Leoše Janáček na scénách Národního divadla v Praze' [Janáček's stage works at the Prague National Theatre], *HRo*, xxxi/4 (1978), 218

S. Štědroňová: 'Leoš Janáček a jeho vztah k Brnu a brněnské univerzitě' [Janáček and his relation to Brno and Brno University], *Universitas: revue brněnské univerzity* (1978), no.4, p.46

J. Mazurek: 'Národopisné výstavy na východní Moravě a ve Slezsku před rokem 1891: podíl Leoše Janáčka na přípravě . . .' [National exhibitions in eastern Moravia and in Silesia before 1891: Janáček's part in their preparation], *Sborník prací pedagogické fakulty v Ostravě*, D15 (1979), 93

V. Gregor: 'L. Janáček a Všeslovanská výstava v Petrohradě roku 1904' [Janáček and the Pan-Slav Exhibition in St Petersburg 1904], *Janáčkiana '78 a '79*, ed. K. Steinmetz (Ostrava, 1980), 124

———: 'L. Janáček a společnost pro výzkum dítěte v Brně' [Janáček and the Society for Child Research in Brno], *OM*, xiii (1981), 265

A. Závodský: 'Petr Bezruč a Leoš Janáček', *Sborník prací filozofické fakulty brněnské univerzity*, D28 (1981), 29

J. Sajner: 'Patografická studie o Leoši Janáčkovi' [A pathological study of Janáček], *OM*, xiv (1982), 233

P. Heyworth: *Otto Klemperer: his Life and Times*, i: *1885–1933* (Cambridge, 1983), 165, 231, 258, 315

Bibliography

S. Přibáňová: 'Nové poznatky k rodokmenu Leoše Janáčka' [New information concerning Janáček's family tree], *Časopis Moravského muzea: vědy společenské*, lxix (1984), 129

JANÁČEK AND OTHER COMPOSERS

J. Racek: 'Janáček a Pavel Křížkovský', *Program [Státního divadla v Brně]*, iii (1948), 371

——: 'Leoš Janáček a Bedřich Smetana', *Slezský sborník*, xlix (1951), 433–84

B. Štědroň: 'Janáček a Čajkovskij', *Sborník prací filosofické fakulty brněnské university*, ii/2–4 (1953), 201 [with Eng. summary]

——: 'Antonín Dvořák a Leoš Janáček', *Musikologie*, v (1958), 105

J. Racek: 'Leoš Janáček o skladebné struktuře klavírních děl Fryderyka Chopina' [Janáček on the structure of Chopin's piano works], *Sborník prací filosofické fakulty brněnské university*, F4 (1960), 5; Fr. trans. in *Annales Chopin*, vi (Warsaw, 1965), 13; Ger. trans. in *Chopin-Jb* [2] (Vienna, 1963), 88, repr. in *Leoš Janáček-Gesellschaft: Mitteilungsblatt 1983*, nos. 1–3

——: 'Leoš Janáčeks und Béla Bartóks Bedeutung in der Weltmusik', *Sborník prací filosofické fakulty brněnské university*, F6 (1962), 5; abridged in *SM*, v (1963), 501

B. Štědroň: 'Leoš Janáček und Ferenc Liszt', *SM*, v (1963), 295; also in *Sborník prací filosofické fakulty brněnské university*, F7 (1963), 139

M. Štědroň: 'Janáček a Schönberg', *Časopis Moravského musea: vědy společenské*, xlix (1964), 237

——: 'Janáček a modernismus třicátých let: Hába, Cowell, Křenek' [Janáček and the modernism of the 1930s], *HRo*, xxiii (1970), 72

B. Štědroň: 'Leoš Janáček a Ludwig van Beethoven', *Universitas*, iii (Brno, 1970), no.4, p.1; Ger. trans. in *Beethoven-Kongress Berlin 1970*, 125

J. Knaus: 'Leoš Janáček und Richard Strauss', *NZM*, Jg.133 (1972), 128, rev. in *Richard-Strauss-Blätter*, new ser., no.3 (Vienna, 1980), 74

A. Gozenpud: 'Janáček a Musorgskij', *OM*, xii (1980), no.4, p.101; no.5, p.1, vii; Ger. trans. in *Leoš Janáček-Gesellschaft: Mitteilungsblatt 1984*

STYLISTIC, ANALYTICAL AND AESTHETIC STUDIES

J. Racek: 'Slovanské prvky v tvorbě Leoše Janáčka' [Slav elements in Janáček's works], *Časopis Matice moravské*, lxx (1951), 364–417; also pubd separately (Brno, 1952)

L. Podéšt': 'O harmonické práci ve sborové tvorbě Leoše Janáčka' [Harmonic practice in Janáček's choral works], *HRo*, x (1957), 137, 189

Janáček

A. Hába: 'Hudební sloh Janáčkův a jeho současníků' [The musical style of Janáček and his contemporaries], *Leoš Janáček a soudobá hudba: Brno 1958*, 117

E. Herzog: 'Harmonie und Tonart bei Janáček', *GfMKB, Köln 1958*, 136

J. Vysloužil: 'Über die Bedeutung der sogenannten modalen Strukturen bei der Entstehung von Janáčeks musikdramatischem Stil', *Operní dílo Leoše Janáčka: Brno 1965*, 32; see also *HRo*, xix (1966), 552

P. Gülke: 'Versuch zur Ästhetik der Musik Leoš Janáčeks', *DJbM*, xii (1967), 5–39; repr. in *Leoš Janáček*, Musik-Konzepte, no.7, ed. H.-K. Metzger and R. Riehn (Munich, 1979), 4–40

M. Štědroň: 'Několik poznámek k Janáčkově tektonice' [Some remarks on Janáček's sense of structure], *Časopis Moravského musea: vědy společenské*, lii (1967), 271 [with Ger. summary]

——: 'Janáček: verismus a impresionismus', *Časopis Moravského musea: vědy společenské*, liii–liv (1968–9), 125–54 [with Ger. summary]

H. Hollander: 'Der Natur-Impressionismus in Janáčeks Musik', *SMz*, cix (1969), 61; also in *Leoš Janáček et musica europaea: Brno III 1968*, 81

J. Knaus: 'Leoš Janáček: Aussenseiter der Neuen Musik', *GfMKB, Bonn 1970*, 458

M. Štědroň: 'Janáček und der Expressionismus', *Sborník prací filosofické fakulty brněnské university*, H5 (1970), 105 [with Cz. summary]; repr. in *Leoš Janáček – Materialien*, ed. J. Knaus (Zurich, 1982), 90

A. Sychra: 'K Janáčkově tematické práci' [Janáček's thematic work], *HRo*, xxiii (1970), 14

J. Tyrrell: 'Janáček and the Speech-melody Myth', *MT*, cxi (1970), 793; Ger. trans., *Leoš Janáček-Gesellschaft: Mitteilungsblatt 1971*, no.2; Cz. version, *OM*, i (1969), 227

O. Chlubna: 'O kompozičním myšlení Leoše Janáčka' [Janáček's compositional thought processes], *HRo*, xxiv (1971), 121

R. Gerlach: 'Leoš Janáček und die Erste und Zweite Wiener Schule: ein Beitrag zur Stilistik seines instrumentalen Spätwerks', *Mf*, xxiv (1971), 119

P. Gülke: 'Protokolle des schöpferischen Prozesses: zur Musik von Leoš Janáček', *NZM*, Jg.134 (1973), 407, 498

N. S. Josephson: 'Formale Strukturen in der Musik Leoš Janáčeks', *Časopis Moravského musea: vědy společenské*, lix (1974), 103

M. Štědroň: 'K podstatě tzv. sociálního a slovanského expresionismu u Leoše Janáčka' [The essence of the so-called social and Slavonic expressionism in Janáček], *Česká hudba světu: svět české hudbě*, ed. J. Bajer (Prague, 1974), 119

Bibliography

B. Štědroň: 'Příznačný motiv u Janáčka' [Leitmotif in Janáček], *HRo*, xxvii (1974), 438

D. Ströbel: *Motiv und Figur in den Kompositionen der Jenufa-Werkgruppe Leoš Janáčeks: Untersuchungen zum Prozess der kompositorischen Individuation bei Janáček* (Munich and Salzburg, 1975)

J. Racek: 'Leoš Janáčeks Kompositionsprinzip in seinen Spätwerk', *Mf*, xxix (1976), 177

B. Štědroň: 'Slavische Motive im Werk Leoš Janáčeks', *Beiträge zur Musikgeschichte Osteuropas*, ed. E. Arro (Wiesbaden, 1977), 284

V. Gregor: 'Zdroje Janáčkova slovanství' [The sources of Janáček's Slavness], *Časopis Slezského muzea*, B27 (1978), 72

V. Hudec: 'Sociální a kulturní fenomény moravského regionu a Janáčkovo dílo' [Social and cultural phenomena of the Moravian region and Janáček's works], *Leoš Janáček ac tempora nostra: Brno XIII 1978*, 99

J. Jiránek: 'Dramatické rysy Janáčkova klavírního stylu' [Dramatic traits in Janáček's writing for the piano], *OM*, x (1978), 139; repr. in J. Jiránek: *Muzikologické etudy* (Prague, 1981), 88

———: 'Janáčkova estetika', *Estetika*, xv (1978), 193; repr. in J. Jiránek: *Muzikologické etudy* (Prague, 1981), 101; abridged Eng. trans. in *MZ*, xvi (1980), 51

R. Pečman: 'Disjunkce estetických názorů a skladebného díla u Leoše Janáčka' [The disparity of Janáček's aesthetic views and compositional practice], *OM*, x (1978), 175

I. Poledňák: 'Logik und Sinn der schöpferischen Persönlichkeit Janáčeks', *Leoš Janáček ac tempora nostra: Brno XIII 1978*, 87; Cz. trans. in *HRo*, xxxi (1978), 520

J. Racek: 'Janáčkův impresionismus', *OM*, x (1978), 161

K. Steinmetz: 'Vliv některých slezských fenomenů na utváření Janáčkova vyzrálého kompozičního stylu sborového' [The influence of certain Silesian phenomena on the shaping of Janáček's style in his mature choral works], *Časopis Slezského muzea*, B27 (1978), 134; see also *HRo*, xxxi (1978), 137, 228

K. Steinmetz and M. Navrátil: 'Janáčkovy "názory o sčasování" a jejich uplatnění ve skladatelově kompoziční praxi' [Janáček's 'views on rhythm' and their part in the composer's compositional practice], *Leoš Janáček ac tempora nostra: Brno XIII 1978*, 195

J. Trojan: 'Co si napisoval Leo Janáček' [How Janáček wrote student exercises], *OM*, x (1978), 166

J. Tyrrell: 'Janáček a viola d'amour', *Leoš Janáček ac tempora nostra: Brno XIII 1978*, 303; enlarged Eng. version in J. Tyrrell: *Leoš Janáček: Kát'a Kabanová* (Cambridge, 1982), 154

J. Volek: 'Leoš Janáček und die neue Art der Auswertung spontaner

Elemente der musikalischen Kreativität im XX. Jahrhundert', *Leoš Janáček ac tempora nostra: Brno XIII 1978*, 41–75

J. Vysloužil: 'Leoš Janáček: úvaha o genezi jeho umělecké osobnosti a slohu' [Reflections on the genesis of Janáček's artistic personality and style], *OM*, x (1978), 131

M. Wehnert: 'Imagination und thematisches Verständnis bei Janáček – dargestellt an "Taras Bulba"', *Leoš Janáček ac tempora nostra: Brno XIII 1978*, 179

Z. Blažek: 'K problému českosti Janáčkovy hudební mluvy' [The problem of the Czechness of Janáček's musical language], *OM*, xi (1979), 263

D. Schnebel: 'Das späte Neue: Versuch über Janáčeks Werke von 1918–1928', *Leoš Janáček*, Musik-Konzepte, no.7, ed. H.-K. Metzger and R. Riehn (Munich, 1979), 74

M. Barvík: 'Uměl Janáček instrumentovat?' [Did Janáček know how to orchestrate?], *Hudební nástroje*, xvii (1980), 24, 59, 102, 141

V. Felix: 'Příspěvek k poznání specifických rysů Janáčkova sonátového slohu: analýza Sonáty pro housle a klavír' [The recognition of specific features of Janáček's sonata style: analysis of the Violin Sonata], *Živá hudba*, vii (1980), 127 [with Ger. summary]

N. S. Josephson: 'Conflicting polarities and their resolution in the music of Leoš Janáček', *Casopis Moravského muzea: vědy společenské*, lxv (1980), 141 [with Cz. summary]

Z. Blažek: 'Janáčkovy skladebné veličiny a skladba komplikační' [Janáček's greatness as a composer and complication-type composition], *OM*, xiii (1981), 65

C. Kohoutek: 'Janáčkův odkaz kompoziční teorii a praxi' [Janáček's legacy to compositional theory and practice], *HRo*, xxxiv (1981), 321

M. Kaňková: 'Sonátová forma v díle Leoše Janáčka' [Sonata form in Janáček's works], *OM*, xiv (1982), 135

J. Vysloužil: 'Eine Skizze zum Vokalstil von A. Dvořák und L. Janáček', *Sborník prací filozofické fakulty brněnské univerzity*, H18 (1983), 7

STAGE WORKS

M. Brod: *Sternenhimmel: Musik- und Theatererlebnisse* (Prague and Munich, 1923, 2/1966 as *Prager Sternenhimmel* . . .; Cz. trans., 1969) [incl. chaps. on *Jenůfa* and *Kát'a Kabanová*]

L. Firkušný: *Odkaz Leoše Janáčka české opeře* [Janáček's legacy to Czech opera] (Brno, 1939) [chaps. on all Janáček's operas]

Musikologie, iii (1955) [incl. F. Pala: 'Jevištní dílo Leoše Janáčka' [Janáček's stage works], 61–210; J. Burjanek: 'Janáčkova Kát'a Kabanová a Ostrovského Bouře' [Janáček's *Kát'a* and Ostrovsky's

Bibliography

The Storm], 345–416; T. Straková: 'Janáčkovy operní náměty a torsa' [Janáček's operatic projects and fragments], 417–49]

T. Straková: 'Janáčkova opera Osud' [Janáček's *Fate*], *Časopis Moravského musea: vědy společenské*, xli (1956), 209–60; xlii (1957), 113–64 [with Ger. summary]

J. Racek: 'Janáček – dramatik', *SH*, ii (1958), 6, 49, 91, 142; abridged Ger. trans. in *DJbM*, v (1960), 37

V. Nosek, ed.: *Opery Leoše Janáčka na brněnské scéně* [Janáček's operas on the Brno stage] (Brno, 1958) [articles on each opera incl. O. Chlubna on his revisions to *Šárka* and *From the House of the Dead*]

F. Pala: 'Postavy a prostředí v Její pastorkyni' [Characters and environment in *Jenůfa*], *Leoš Janáček: sborník statí a studií* (Prague, 1959), 29–70

B. Štědroň: 'K Janáčkově opeře Osud' [Janáček's *Fate*], *Živá hudba*, i (1959), 159–83

K. Wörner: 'Katjas Tod: die Schlussszene der Oper "Katja Kabanova" von Janáček', *SMz*, xcix (1959), 91; also in *Leoš Janáček a soudobá hudba: Brno 1958*, 392 [see also K. H. Wörner: 'Natur, Liebe und Tod bei Janáček', *Die Musik in der Geistegeschichte* (Bonn, 1970), 131; 'Leoš Janáček', *Das Zeitalter der thematischen Prozesse in der Geschichte der Musik* (Regensburg, 1969), 146; Eng. trans. in J. Tyrrell: *Leoš Janáček: Kát'a Kabanová* (Cambridge, 1982), 174]

M. Očadlík: 'Dvě kapitoly k Janáčkovým Výletům pana Broučka na měsíc' [Two chapters on Janáček's *Mr Brouček's Excursion to the Moon*], *MMC* (1960), no.12, pp.133–47

Z. Sádecký: 'Celotónový charakter hudební řeči v Janáčkově "Lišce Bystroušce"' [The whole-tone character of the musical language of Janáček's *Cunning Little Vixen*], *Živá hudba*, ii (1962), 95–163 [with Ger. summary]

A. Závodský: *Gabriela Preissová* (Prague, 1962) [incl. chap. on the play on which Janáček based *Jenůfa*]

O. Fiala: 'Libreto k Janáčkově opeře Počátek románu' [The libretto of Janáček's *The Beginning of a Romance*], *Časopis Moravského musea: vědy společenské*, xlix (1964), 192–222 [with Ger. summary]

T. Straková: 'Mezihry v Káti Kabanové' [The interludes in *Kát'a Kabanová*], *Časopis Moravského musea: vědy společenské*, xlix (1964), 229; Ger. trans. in *Operní dílo Leoše Janáčka: Brno 1965*, 39, repr. in *Leoš Janáček – Materialien*, ed. J. Knaus (Zurich, 1982), 134; abridged Eng. trans. in J. Tyrrell: *Leoš Janáček: Kát'a Kabanová* (Cambridge, 1982), 134

J. Procházka: 'Brod's Übersetzung des Librettos der Jenůfa und die Korrekturen Franz Kafkas', *Operní dílo Leoše Janáčka: Brno 1965*,

109, repr. in *Leoš Janáček – Materialien*, ed. J. Knaus (Zurich, 1982), 30

———: 'Z mrtvého doma: Janáčkův tvůrčí i lidský epilog a manifest' [*From the House of the Dead*: Janáček's creative and human epilogue and manifesto], *HV*, iii (1966), 218–43, 426–37

J. Tyrrell: 'The Musical Prehistory of Janáček's Počátek románu and its Importance in Shaping the Composer's Dramatic Style', *Časopis Moravského musea: vědy společenské*, lii (1967), 245–70 [with Cz. summary]

G. Abraham: 'Realism in Janáček's Operas', *Slavonic and Romantic Music* (London, 1968), 83

A. Mazlová: 'Zeyerova a Janáčkova Šárka' [Zeyer's and Janáček's *Šárka*], *Časopis Moravského musea: vědy společenské*, liii–liv (1968–9), 71 [with Eng. summary]

L. Polyakova: *Opernoye tvorchestvo Leosha Yanacheka* [Janáček's operas] (Moscow, 1968)

Z. Sádecký: 'Výstava dialogu a monologu v Janáčkově Její pastorkyni' [Dialogue and monologue structure in Janáček's *Jenůfa*], *Živá hudba*, iv (1968), 73–146 [with Ger. summary]

B. Štědroň: *Zur Genesis von Leoš Janáčeks Oper Jenůfa* (Brno, 1968, rev. 2/1972); extracts in Eng. in *Sborník prací filosofické fakulty brněnské university*, H3 (1968), 43–74; H5 (1970), 91 [incl. extensive bibliography]

J. Tyrrell: 'Mr Brouček's. Excursion to the Moon', *Časopis Moravského musea: vědy společenské*, liii–liv (1968–9), 89–124 [with Cz. summary]

K. von Fischer: 'Zu Leoš Janáčeks "Das Schlaue Füchslein" ', *Jb 70/71 Opernhaus Zürich*, xlix (1970), 10; repr. in *Leoš Janáček-Gesellschaft: Mitteilungsblatt 1972*, no.3

L. Polyakova: 'Russkiye operï Yanacheka' [Janáček's Russian operas], *Puti razvitiya i vzaimosvyazi russkovo i chekhoslovatskovo*, ed. Institut istorii iskusstv (Moscow, 1970), 190; Cz. trans. in *Cesty rozvoje a vzájemné vztahy ruského a československého umění* (Prague, 1974), 247

E. Chisholm: *The Operas of Leoš Janáček* (Oxford, 1971)

T. Kneif: *Die Bühnenwerke von Leoš Janáček* (Vienna, 1974)

A. Němcová: 'Brněnská premiéra Janáčkovy Její pastorkyně' [The Brno première of Janáček's *Jenůfa*], *Časopis Moravského musea: vědy společenské*, lix (1974), 133 [with Eng. summary]; Ger. trans. in *Leoš Janáček – Materialien*, ed. J. Knaus (Zurich, 1982), 7

T. Straková: 'Janáčkova opera Její pastorkyňa: pokus o analýzu díla' [Janáček's *Jenůfa*: an attempt at an analysis], *Časopis Moravského musea: vědy společenské*, lix (1974), 119 [with Ger. summary]

F. Pulcini: 'Le opere teatrali inedite di Leoš Janáček', *NRMI*, ix (1975), 552

Bibliography

M. Štědroň: 'K analýze vokální melodiky Janáčkovy opery Věc Makropulos s využitím samočinného počitače' [The analysis of the voice part of Janáček's opera *The Makropulos Affair* with the aid of a computer], *HV*, xii (1975), 46 [with Ger. summary]; see also *Musicologica slovaca*, vi (1978), 187

M. Ewans: *Janáček's Tragic Operas* (London, 1977; Ger. trans., 1981)

J. Burghauser: 'Janáčkovo poslední hudebně dramatické torso' [Janáček's last musico-dramatic fragment], *HV*, xv (1978), 317 [incidental music to *Schluk und Jau*]

F. Pala: 'Osud', *HRo*, xxxi (1978), 41; see also *HRo*, xxxi/5 (1978), 231

R. Pečman: 'Námětova geneze Věci Makropulos (Cesta od Shawa k Čapkovi a Janáčkovi)' [The genesis of the subject of *The Makropulos Affair*: the journey from Shaw to Čapek and Janáček], *Leoš Janáček ac tempora nostra: Brno XIII 1978*, 245; abridged Ger. trans. in *Sborník prací filozofické fakulty brněnské univerzity*, H17 (1982), 21

L. Peduzzi: 'Janáček, Haas a Divoška' [Janáček, Haas and *The Tomboy*], *OM*, x (1978), no.8, pp.i–iv; xii (1980), no.7, pp.i–iv, viii

L. Polyakova: *Cheshskaya i slovatskaya opera XX. veka* [Czech and Slovak opera of the 20th century], i (Moscow, 1978)

J. Tyrrell: disc notes to Decca recordings of Janáček operas conducted by Charles Mackerras: *Věc Makropulos*, D144D 2 (1979); *Z mrtvého domu*, D224D 2 (1980); *Příhody Lišky Bystroušky*, D257D 2 (1982); *Jenůfa*, D276D 3 (1983)

A. Němcová, S. Přibáňová and T. Straková: surveys of source materials for all the operas, *Časopis Moravského muzea: vědy společenské*, lxv (1980), 149

J. Tyrrell: 'Mr Brouček at Home: an Epilogue to Janáček's Opera', *MT*, cxx (1980), 30 [on the discarded act of the *Excursion to the Moon*]

A. Gozenpud: *Dostoevskiy i muzïkal'no-teatral'noye iskusstvo* [Dostoyevsky and the art of music-theatre] (Leningrad, 1981), 202

D. Plamenac: 'Nepoznati komentari Leoša Janáčeka operi "Katja Kabanova" ' [An unknown commentary by Janáček on his opera *Kát'a Kabanová*], *MZ*, xvii (1981), 122

F. Pulcini: *Per una ricognizione italiana del teatro musicale di Leoš Janáček: aspetti della drammaturgia di 'Da una casa di morti'* (diss., U. of Turin, 1981–2)

J. Tyrrell: *Leoš Janáček: Kát'a Kabanová* (Cambridge, 1982)

VOCAL

Č. Gardavský: 'Chrámové a varhanní skladby Leoše Janáčka' [Janáček's church and organ compositions], *Musikologie*, iii (1955), 330

Janáček

F. Trávníček, B. Štědroň and L. Kundera: *Zápisník zmizelého* (Brno, 1956)

R. Večerka: 'K historii textu Janáčkovy Hlaholské mše' [The history of the text of Janáček's *Glagolitic Mass*], *Sborník prací filosofické fakulty brněnské university*, F1 (1957), 64

J. Šmolík: 'Sbory Leoše Janáčka na texty Petra Bezruče' [Janáček's choruses on Bezruč texts], *Leoš Janáček: sborník statí a studií* (Prague, 1959), 73–112

B. Štědroň: 'Předzvěsti Janáčkovy opery Její pastorkyňa' [Precursors of Janáček's opera *Jenůfa*], *Časopis Moravského musea: vědy společenské*, li (1966), 291; Ger. trans. in B. Štědroň: *Zur Genesis von Leoš Janáčeks Oper Jenůfa* (Brno, rev. 2/1972), 11 [on the chorus *Žárlivec* – The Jealous Man]

V. Telec: 'Janáčkův smuteční sbor Odpočiň si! [Janáček's funeral chorus *Take your rest!*], *HV*, iii (1966), 158

K. Svoboda: 'K autorství Zápisníku zmizelého' [On the authorship of *The Diary of one who Disappeared*], *Vlastivědný věstník moravský*, xxii (1970), 211

A. Tučapský: *Mužské sbory Leoše Janáčka a jejich interpretační tradice* [Janáček's male-voice choruses and their performance tradition] (Prague, 1971)

Z. Blažek: 'Osamělá bez těchy: prvotní neznámé znění Janáčkova sboru' [The unknown original version of Janáček's chorus *Alone without comfort*], *OM*, x (1978), 101; see also *OM*, x (1978), no.9, pp.i–ii and *OM*, xi (1979), no.1, pp. ii–iv

M. Kaduch: ' "Ohlas" jako historický a umělecký jev: na okraj Janáčkovy sborové tvorby komponované v letech 1873 až 1914 na texty lidových písní' ['Echoes [of folksong]' as a historical and artistic phenomenon: on the margin of Janáček's choral works 1873–1914 on folksong texts], *OM*, x (1978), 168; see also *OM*, xiv (1982), 44

INSTRUMENTAL

Musikologie, iii (1955) [incl. J. Burghauser: 'Janáčkova tvorba komorní a symfonická' [Janáček's chamber and symphonic works], 211–305; L. Kundera: 'Janáčkova tvorba klavírní' [Janáček's piano works], 306]

T. Straková: 'Neznáme nástrojové skladby Leoše Janáčka' [Janáček's unknown instrumental works], *Časopis Moravského musea: vědy společenské*, xliv (1959), 163 [with Ger. summary]

B. Heran: 'Po stopách Janáčkovy Pohádky' [On the track of Janáček's *Fairy tale*], *HRo*, xvii (1964), 872

S. Přibáňová: 'K otázce vzniku Janáčkova Tarase Bulby' [The origin of Janáček's *Taras Bulba*], *Časopis Moravského musea: vědy společenské*, xlix (1964), 223 [with Ger. summary]

Bibliography

J. D. Link: 'Janáčeks Ouvertüre Žárlivost und ihre Beziehung zur Oper Jenufa', *Operní dílo Leoše Janáčka: Brno 1965*, 131

H. Hollander: 'Die thematischen Metamorphosen in Janáčeks zweitem Streichquartett', *NZM*, Jg.127 (1966), 95

A. Němcová: 'Janáčkova houslová sonáta: geneze díla' [The genesis of Janáček's Violin Sonata], *Časopis Moravského musea: vědy společenské*, lii (1967), 289 [with Ger. summary]

B. Štědroň: 'Precursors of Janáček's Opera "Její pastorkyňa" (Jenůfa): Prologue to Jenůfa – Jealousy', *Sborník prací filosofické fakulty brněnské university*, H3 (1968), 43–73; Ger. trans. in B. Štědroň: *Zur Genesis von Leoš Janáčeks Oper Jenůfa* (Brno, 1968, enlarged 2/1972), 21

M. Wehnert: 'Zur Ästhetik der motivisch-thematischen Gestaltung der Sinfonietta von Leoš Janáček', *Leoš Jaňáček et musica europaea: Brno III 1968*, 171

K. Janeček: 'O kompozičním myšlení Leoše Janáčka: úvaha nad partiturou Symfonietty' [Janáček's compositional thought: a reflection on the score of the Sinfonietta], *HRo*, xxiii (1969), 315

Z. Sádecký: 'Tematické vztahy v Janáčkově klavírním díle' [Thematic relations in Janáček's piano works], *HV*, vi (1969), 26–56 [with Eng. and Ger. summaries]

J. Ludová: 'O otázce kvantitativních analytických metod' [The question of quantitative analytical methods], *HV*, viii (1971), 183 [analysis of *On the Overgrown Path*]

B. Štědroň: 'Die Inspirationsquellen von Janáčeks Concertino', *Musica cameralis: Brno VI 1971*, 423; Cz. orig. in *Na křižovatce umění: sborník k poctě . . . Artura Závodského* (Brno, 1973), 363

H. Hollander: 'Der Klassizismus in Janáčeks Sinfonietta', *NZM*, Jg.135 (1974), 685

J. Uhde: 'Ein musikalisches Monument (zu Leoš Janáčeks Klaviersonaten-Fragment "1.10.1905" ', *Zeitschrift für Musiktheorie*, vi/2 (1975), 89; repr. in *Leoš Janáček – Materialien*, ed. J. Knaus (Zurich, 1982), 81

P. Andraschke: 'Analytische Beobachtungen am ersten Satz der Violinsonate von Leoš Janáček', *Leoš Janáček ac tempora nostra: Brno XIII 1978*, 221

T. Hirsbrunner: 'Absolut-musikalische und musikdramatische Aspekte in Janáčeks erstem Streichquartett', *Leoš Janáček ac tempora nostra: Brno XIII 1978*, 273

A. Gozenpud: 'K otázce myšlenkových shod Tarase Bulby a Výletu pana Broučka do XV. století' [The coincidence of ideas behind *Taras Bulba* and *Mr Brouček's Excursion to the 15th Century*], *Leoš Janáček ac tempora nostra: Brno XIII 1978*, 235

V. Lébl: 'Janáčkova Sinfonietta', *HV*, xv (1978), 305

73

M. Štědroň: 'Janáčkova symfonie Dunaj v pojetí Osvalda Chlubny' [Janáček's symphony *The Danube* as understood by Osvald Chlubna], *HV*, xv (1978), 326

D. Holland: 'Kompositionsbegriff und Motivtechnik in Janáčeks Streichquartetten', *Leoš Janáček*, Musik-Konzepte, no.7, ed. H.-K. Metzger and R. Riehn (Munich, 1979), 67

J. Vičar: 'K problematice 1. věty (Smrt Andrijova) Janáčkova Tarase Bulby' [The problem of the first movement (the death of Andriy) of Janáček's *Taras Bulba*], *Janáčkiana '78 a '79*, ed. K. Steinmetz (Ostrava, 1980), 127

M. Štědroň: 'Die Bassklarinette in Janáčeks Sextett "Jugend"', *Sborník prací filozofické fakulty brněnské univerzity*, H18 (1983), 25

THEORETICAL AND PEDAGOGICAL

O. Chlubna: 'Janáček – učitel' [Janáček as teacher], *Musikologie*, iii (1955), 51

M. Černohorská: 'K problematice vzniku Janáčkovy teorie nápěvků' [Problems about the origin of Janáček's speech-melody theory], *Časopis Moravského musea: vědy společenské*, xlii (1957), 165; see also xliii (1958), 129 [both with Ger. summaries]

J. Volek: 'Janáčkova Uplná nauka o harmonii' [Janáček's *Complete Harmony Manual*], J. Volek: *Novodobé harmonické systémy z hlediska vědecké filosofie* (Prague, 1961), 231–78

Z. Blažek: 'Janáčkova hudební teorie', *Leoš Janáček: hudební teoretické dílo*, ed. Z. Blažek, i (Prague, 1968), 21 [further bibliography in J. Racek's introduction, 9]

B. Dušek: 'Janáčkovy nazory na hudební harmonii v letech 1884–1912' [Janáček's views on harmony in music in the years 1884 to 1912], *HV*, v (1968), 374–404 [with Eng., Ger. and Russ. summaries]

J. Blatný: 'Janáček učitel a teoretik' [Janáček as teacher and theorist], *OM*, i (1969), 97

Z. Blažek: 'Polyphonie und Rhythmik in Janáčeks Musiktheorie', *Sborník prací filosofické fakulty brněnské university*, H4 (1969), 107 [with Cz. summary]; Cz. orig., *OM*, xii (1980), no.2, p.v

F. Řehánek: 'Ke spojovacím formám a spojům v Janáčkově učení o harmonii' [Forms of combination and links in Janáček's theories of harmony], *HV*, vi (1969), 439

——: 'Janáčeks Beitrag zur Harmonielehre', *Musica bohemica et europaea: Brno V 1970*, 435

——: 'Janáček a funkčnost v jeho teorii' [Janáček and harmonic function in his theories], *Časopis Moravského musea: vědy společenské*, lix (1974), 97

Bibliography

——: 'K Janáčkově terminologii' [Janáček's terminology], *OM*, viii (1976), 108

Articles on Janáček's teaching activities by M. Vysloužilová, J. Fukač and O. Settari, *OM*, viii (1976), 214

F. Řehánek: 'Janáček a tonalita', *Leoš Janáček ac tempora nostra: Brno XIII 1978*, 279

B. Küfhaberová: 'Janáčkova hudební terminologie' [Janáček's musical terminology], *Janáčkiana '78 a '79*, ed. K. Steinmetz (Ostrava, 1980), 110; see also *Leoš Janáček ac tempora nostra: Brno XIII 1978*, 289

K. Steinmetz: 'Janáčkova nauka o formaci, útvarnosti skladeb' [Janáček's teaching on form in composition], *OM*, xii (1980), 273

M. Beckerman: 'Janáček and the Herbartians', *MQ*, lxix (1983), 388

FOLK MUSIC

J. Procházka: *Lašské kořeny života i díla Leoše Janáčka* [The Lašsko roots in Janáček's life and works] (Prague, 1948)

J. Vysloužil: 'Hudebně folkloristické dílo Leoše Janáčka' [Janáček's work on folk music], *Leoš Janáček: O lidové písni a lidové hudbě*, ed. J. Vysloužil (Prague, 1955), 29–78

B. Štědroň: 'K Janáčkovým národním tancům na Moravě' [Janáček's *National Dances of Moravia*], *Sborník prací filosofické fakulty brněnské university*, F2 (1958), 44 [with Ger. summary]; see also *OM*, vi (1974), 165

J. Vysloužil: 'Janáčkova tvorba ve světle jeho hudebně folkloristické teorie' [Janáček's works in the light of his folk music theories], *Sborník Janáčkovy akademie múzických umění*, ii (1960), 53

K. Vetterl: 'Lidová píseň v Janáčkových sborech do roku 1885' [Folksong in Janáček's choruses to 1885], *Sborník prací filosofické fakulty brněnské university*, F9 (1965), 365 [with Ger. summary]; see also 'Janáček's Creative Relationship to Folk Music', *Leoš Janáček et musica europaea: Brno III 1968*, 235

H. Malá: 'Statistická srovnávací analýza sborů Leoše Janáčka a moravských lidových písní' [A statistical comparative analysis of Janáček's choruses and Moravian folksongs], *HV*, iv (1967), 602 [with Eng. and Ger. summaries]

M. Malura: 'Rozbor Janáčkových Lašských tanců z hlediska folklórní provenience tematického materiálu' [Analysis of Janáček's *Dances from Lašsko* based on the provenance of the themes], *Časopis Slezského muzea: vědy společenské*, xx (1971), ser.B, 22

J. Vysloužil: 'Lidová písňová tradice v teorii a díle Leoše Janáčka' [Folksong tradition in Janáček's theories and works], *Národopisné aktuality*, viii (1971), no.1, p.3

A. Geck: *Das Volksliedmaterial Leoš Janáčeks: Analysen der Struk-*

75

turen unter Einbeziehung von Janáčeks Randbemerkungen und Volkstudien (Regensburg, 1975)

F. Hrabal and O. Hrabalová: 'K některých Janáčkovým úpravám lidových písní z let 1908–1912' [Janáček's folksong arrangements 1908–12], *Časopis Slezského muzea: vědy společenské*, xxv (1976), ser.B, 14

D. Holý: 'Janáčkovo pojetí lidové písně a hudby' [Janáček's concept of folksong and folk music], *Leoš Janáček ac tempora nostra: Brno XIII 1978*, 105; also in *Národopisné aktuality*, xv (1978), 271

Z. Mišurec: 'Činnost Leoše Janáčka jako organizátora národopisné práce' [Janáček's activities in the organization of ethnographic work], *Český lid*, lxv (1978), 221 [with Ger. summary]

M. Malura: 'Neznámé prameny ke vztahu L. Janáčka k regionu' [Unknown sources on Janáček's regional connections], *Janáčkiana '78 a '79*, ed. K. Steinmetz (Ostrava, 1980), 119

J. Trojan: 'Janáčkovy klavírní doprovody Moravských lidových písní' [Janáček's piano accompaniments to his Moravian Folksongs], *Živý Janáček*, ed. Z. Zouhar (Brno, 1980), 25

——: *Moravská lidová píseň* [Moravian folksong] (Prague, 1980)

JANÁČEK AS WRITER

A. Novák: 'Leoš Janáček spisovatel' [Janáček as writer]; V. Helfert: 'Kořeny Janáčkova kritického stylu' [The roots of Janáček's critical style], J. Racek and L. Firkušný: *Janáčkovy feuilletony z L.N.* (Brno, 1938), 15, 22; repr. in L. Janáček: *Fejetony z Lidových novin* (Brno, 1958), 265, 273

P. Eisner: 'Janáček spisovatel' [Janáček as writer], *HRo*, xi (1958), 762

A. Sychra: 'Janáčkův spisovatelský sloh, klíč sémantice jeho hudby' [Janáček's style as a writer: a key to the semantics of his music], *Estetika*, i (1964), 3–30, 109–25

J. Kvapil: 'Janáček skladatel a beletrista' [Janáček as composer and writer], *OM*, xi (1979), 134

——: 'K problému Janáčkova literárního slohu' [The problem of Janáček's literary style], *Janáčkiana '78 a '79*, ed. K. Steinmetz (Ostrava, 1980), 102

E. Horová: 'Vztah Leoše Janáčka k jazyku' [Janáček's relation to language], *Živý Janáček*, ed. Z. Zouhar (Brno, 1980), 15

'Janáček a literatura' [articles by T. Straková, A. Němcová, K. Steinmetz and S. Přibáňova], *Hudba a literatura*, ed. R. Pečman (Frýdek-Mistek, 1983), 98

PERFORMANCE, INTERPRETATION AND RECEPTION

J. Telcová: articles on the scenography of Janáček's operas in *Časopis*

Bibliography

Moravského musea: vědy společenské, xlix (1964), 259 [on *Jenůfa*]; ibid, 1 (1965), 261 [on the *Vixen*]; ibid, li (1966) 345 [*Kát'a Kabanová*]; lii (1967), 315 [*Brouček*]; liii–iv (1968–9), 155 [*Věc Makropulos*]; lvi (1971), 153 [*Šárka* etc] [all with Ger. summaries and illustrations]; general survey in *Otázky divadla a filmu*, iii (Brno, 1973), 39 [with Eng. summary]

S. Přibáňová: 'Janáčkovy opery ve světle zahraničních kritik' [Janáček's operas in the light of foreign criticism], *Časopis Moravského musea: vědy společenské*, 1 (1965), 231 [with Ger. summary]; see also *Janáčkiana '78 a '79*, ed. K. Steinmetz (Ostrava, 1980), 25

Operní dílo Leoše Janáčka: Brno 1965 [incl. sections on interpretation and staging, notably articles by J. Vogel, p.145 and C. Mackerras, p.102; further articles by Vogel in *OM*, ii (1970), 70 and vii (1975), 229, and final chapter of his biography, rev. 2/1981, p.385; further article by Mackerras in J. Tyrrell: *Leoš Janáček: Kát'a Kabanová* (Cambridge, 1982), 143]

Sborník Janáčkovy akademie múzických umění, v (1965) [devoted to questions of performance and interpretation of Janáček's works]

W. Felsenstein: 'Vier Briefe zum "Schlauen Füchslein" ', W. Felsenstein: *Schriften zum Musiktheater* (Berlin, 1976), 318; repr. in *Leoš Janáček –Materialien*, ed. J. Knaus (Zurich, 1982), 50

M. Kuna: 'Interpretace Janáčkova I. smyčcového kvarteta' [The interpretation of Janáček's first string quartet], *HV*, xiv (1977), 99–144; xv (1978), 340–57; xvi (1979), 43–65 [quantitative analyses of performances of first two movts.]

Leoš Janáček ac tempora nostra: Brno XIII 1978 [incl. section on reception history]

Živý Janáček, ed. Z. Zouhar (Brno, 1980) [incl. several articles on the interpretation of Janáček's works]

S. Přibáňová, A. Simpson and B. Hampton Renton: 'Stage history and reception', J. Tyrrell: *Leoš Janáček: Kát'a Kabanová* (Cambridge, 1982), 111

GUSTAV MAHLER

Paul Banks

Donald Mitchell

CHAPTER ONE

Childhood, early works, 1860–80

Gustav Mahler was born in Kalischt, near Iglau (now Kalištĕ, Jihlava), Bohemia, on 7 July 1860. Of Jewish parents, he was the second of 14 children and the first of the six who survived to maturity. His father, Bernhard, was a man of humble origin, who by sheer determination and ruthlessness established a successful business consisting of a distillery and several taverns in Iglau, where the family moved in autumn 1860. Iglau provided an environment that included military, popular, folk and art music from which Mahler later drew some elements of his musical vocabulary. As a child he learnt a large repertory of folksongs, and received tuition in the piano and theory from several local musicians under whose guidance he rapidly developed into an able pianist and gave his first public concert in 1870. The following year, in an attempt to improve his general academic performance, he was sent to Prague where he attended the Neustädter Gymnasium, but the move was not successful and he returned to Iglau in 1872. Continuing his musical studies, he appeared in public on three further occasions and worked on a (lost) opera, *Herzog Ernst von Schwaben*.

In 1875 Mahler's talent attracted the attention of a farm manager, who persuaded Bernhard that his son should receive a thorough musical training in Vienna, and after an interview with Julius Epstein that September, Mahler was accepted as a student at the conservatory. There he studied the piano with Epstein (1875–7), harmony with Robert Fuchs (1875–6) and composition with Franz Krenn (1875–8). Although successful as a pianist in the 1876 and 1877 conservatory competitions, Mahler abandoned the instrument in favour of composition, in which he graduated in 1878 with the submission of a scherzo for piano quintet (lost). Nevertheless, the experience of playing the piano music of Beethoven, Chopin, Schubert and Schumann left a marked impression on his own music. The composition course was of dual importance to his subsequent career, for, apart from the usual tuition, students were also expected to conduct rehearsals of, and in some cases public performances by, the student orchestra.

Mahler's first attempts at composition date from his childhood years at Iglau, but the earliest surviving works were written during his period at the conservatory, the influence of which is reflected in the rigorous motivic treatment in the movement for piano quartet (?1876–8). He drew some inspiration and stimulus from a wide range of other composers' music, from predecessors, including Beethoven, Mendelssohn, Schubert, Schumann, Wagner, Weber and, increasingly as he grew older, J. S. Bach, as well as from at least one contemporary, Richard Strauss. He had admired Wagner's music before he arrived in Vienna, and it was that interest, coupled with attendance at Krenn's composition classes, which probably formed the basis of early friendships with Hugo Wolf, Rudolf Krzyzanowski, Hans Rott and Anton Krisper; it was an interest in Wagner's political and philosophical ideas and in the early works of

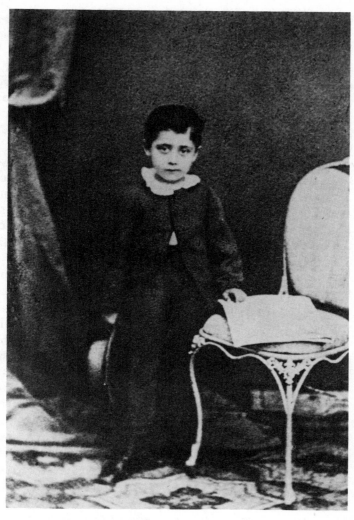

5. *Gustav Mahler aged five or six*

Nietzsche that drew him into a circle that included the future socialist politicians Victor Adler and Engelbert Pernerstorfer.

In 1877 and again in 1878 and 1880 Mahler enrolled as a student at the University of Vienna for a wide range of historical and philosophical subjects indicative of the scope of his lifelong intellectual interests; he also attended some of Bruckner's lectures, though in later years he stated categorically that he was not Bruckner's pupil. In 1877 he prepared and in 1880 published (possibly with Krzyzanowski's help) a piano-duet arrangement of Bruckner's Third Symphony and in 1910 donated the profits from the sale of his own works to the promotion of Bruckner's scores; however, as a conductor he used heavily cut versions of the Fifth and Sixth Symphonies.

After leaving the conservatory in 1878 Mahler attended university lectures, worked as a music teacher and during the years of uncertainty immediately before the beginning of his career as a conductor, composed his first important work, *Das klagende Lied*. The text of the cantata was completed in March 1878 and the composition of the music spread over the following two years. In its original form in three parts, it was sent to Liszt (who disliked the text) and was entered unsuccessfully in 1881 for the Beethoven Prize. In 1892–3 Mahler revised the score, omitting the first part (entitled *Waldmärchen*), touching up the orchestration and deleting the offstage orchestra (the latter was restored in 1898–9 when the work was revised again for publication). Even before the original version was complete Mahler began work on an opera, *Rübezahl*, writing a libretto, in five acts with a prologue, which has survived.

He discussed the idea of a fairy-tale opera with Hugo Wolf, who subsequently felt that Mahler had stolen the idea for the *Rübezahl* text from him: as a result of this misunderstanding their friendship came to an end. The libretto obviously retained its attraction for Mahler, and it seems probable that some music for the opera was composed in the 1880s. Related to this work and to *Das klagende Lied* are the three songs which Mahler composed in 1880 during a youthful romance. The last of them uses a text (and possibly music) from *Rübezahl*, the whole piece with a few alterations eventually becoming *Hans und Grethe* in the first volume of *Lieder und Gesänge*; the first two songs use melodic material from the cantata.

CHAPTER TWO

Bad Hall, Laibach, Olmütz, Kassel, 1880–85

In early summer 1880, on the advice of the publisher Rättig, Mahler found an agent and was offered a conducting post at a small summer theatre at Bad Hall in Upper Austria. The company was small, the artistic standards were low, and the repertory consisted entirely of operetta; but having taken up the baton, Mahler was unable to lay it down, and he asked his agent to find a similar appointment for the winter season. In the absence of such a post, he stayed in Vienna (where he completed *Das klagende Lied*) until the following summer, when he was engaged as Kapellmeister at the Landestheater in Laibach (now Ljubljana). Though small, this theatre offered much better facilities than those at Bad Hall as well as a more interesting repertory, and it was there that Mahler conducted his first opera, *Il trovatore* (3 October 1881). Even at such an early stage of his career, his search for the technically flawless presentation of opera was recognized by local critics. But once again at the end of the season he was without a post, and so he returned to Vienna in early summer 1882. He devoted himself to composition, working on *Rübezahl* and possibly the Richard Leander settings, until an unexpected telegram summoning him to the Stadttheater in Olmütz (now Olomouc) arrived on 10 January 1883. Mahler's appointment was occasioned by the sudden departure of the previous Kapellmeister. The theatre had a larger company than Laibach's, but was in a depressed financial and artistic condition; it was a challenging situation for an aspiring conductor. The local press took an unfavourable view of the appointment, and Mahler's authoritarian attitude alienated some of the singers, but eventually the discipline he imposed was reflected in the improved standards of performance. Although the achievement was only modest, and Mahler was glad to have avoided mounting productions of Mozart and Wagner under such conditions, his work with the provincial company attracted the attention of Karl Ueberhorst, a producer at Dresden, who later recommended him to the Intendant at Kassel.

Back in Vienna, Mahler took up a temporary post as chorus master for a season of Italian opera at the Carltheater, and in May negotiations for an appointment as music and chorus director at Kassel began. Success at Olmütz had paved the way for this new appointment, which depended on a week's trial. Having demonstrated his competence to the satisfaction of the Intendant, Mahler returned to Iglau and Vienna for the summer, making a short trip to Bayreuth – his first – to see *Parsifal*.

Soon after his return Mahler discovered the unsatisfactory working conditions at Kassel. His superior, the Kapellmeister Wilhelm Treiber, was unsympathetic, the regulations governing his activity restrictive, and his repertory was almost entirely light opera. Some compensations were to be found in the performances at Kassel of the Meiningen Hofkapelle under Bülow, which inspired Mahler's intense admiration, and the opportunity to conduct one of his favourite operas, *Der*

6. *Mahler conducting: silhouette by Hans Böhler*

Freischütz, for the first time (10 January 1884). The critical reception of Mahler's conducting was mixed: in general the precision of ensemble and his enthusiasm were praised, but his excessive bodily movements and unconventional tempos met with criticism. Throughout his career Mahler was berated for his unconventional interpretations, and it was at the mindless routine which results from an unthinking acceptance of convention that his best-known dictum – 'Tradition ist Schlamperei' – was aimed.

In 1883–4 Mahler fell in love with one of the singers at Kassel, Johanna Richter. This unhappy affair led to the composition, probably between December 1884 and January 1885, of his first masterpiece, the song cycle *Lieder eines fahrenden Gesellen*, and the inception of the closely related First Symphony. As in *Das klagende Lied* and *Rübezahl*, the texts were written by Mahler himself. Although the songs were intended for voice and orchestra from the first, the orchestration was delayed until 1891–3 and the score was further revised before the first performance in 1896. Because of Mahler's heavy conducting schedule, work on the piece was frequently interrupted, and he also had to compose an occasional work, incidental music to Scheffel's *Der Trompeter von Säkkingen* (lost), for performance at a charity concert. The music was successful (despite Mahler's later disparagement of it), and one movement was included as 'Blumine' in the original version of the First Symphony.

In December 1884 Mahler contacted Neumann, director of the Deutsches Landestheater (German Opera) in Prague, and then Pollini in Hamburg and Staegemann in Leipzig, in the hope of securing a new post. By the end of the following January he had been offered a post as a second conductor at Leipzig, to begin in July 1886. In February 1885 at Münden, near Kassel, he conducted an oratorio (*The Seasons*) for the first time, and largely because of this success was asked to conduct at the summer festival there. Since the Kassel theatre orchestra was to be used it was expected that Kapellmeister Treiber would conduct; because of Mahler's engagement the orchestra, out of loyalty to Treiber, refused to appear, and Mahler's position at the theatre became increasingly difficult. The situation was not improved by Mahler's frequent absences at rehearsals for the festival performances of Mendelssohn's *St Paul*, and he repeatedly came into conflict with the administration over matters of discipline. Eventually his requests for the termination of his contract were agreed to, and the date fixed as 1 July 1885. For a time he faced the prospect of a year without a post, but eventually he was engaged by Neumann as first conductor at Prague, an appointment that made him regret his agreement with Staegemann. Much of his remaining time at Kassel was spent in rehearsing the choirs for *St Paul*, and his efforts were rewarded on 29 June when the oratorio was given before an enthusiastic audience.

CHAPTER THREE

Prague and Leipzig, 1885–8

Mahler's arrival in Prague in July 1885 coincided with that of Angelo Neumann, who had been engaged to rescue the Deutsches Landestheater from financial and artistic ruin. Mahler was originally engaged together with two other conductors, Anton Seidl and Ludwig Slansky. Seidl left after only one month and it was expected that Slansky, as the senior, would take over as chief conductor. Neumann's decision that the two remaining conductors would share the repertory naturally aroused bitterness among Slansky's admirers in the press, orchestra and city. During this year at Prague, Mahler for the first time conducted the stage works of the composers with whom he was to be most closely associated for the rest of his career as an opera conductor: by Mozart (*Don Giovanni, Die Entführung aus dem Serail, Così fan tutte*), Wagner (*Tannhäuser, Die Meistersinger, Das Rheingold, Die Walküre*), Gluck (*Iphigénie en Aulide*) and Beethoven. Mahler's conducting was praised and criticized in much the same terms as at Kassel, but he had become increasingly concerned with both musical and dramatic faithfulness to the score. Towards the end of the season he also took part in two concerts, the second of which included the first public performance of at least one of his songs. The last months at Prague were marred by disagreements between Neumann the businessman and

Mahler the artist, but Mahler nevertheless tried, vainly, to have the hastily accepted contract with Staegemann, due to begin in July, annulled. After the close of the Prague season he spent a brief holiday with his ailing parents, and then in July made his way to Leipzig.

The Neues Stadttheater, Leipzig, was better equipped than any that had previously employed Mahler, having a large orchestra led by Henri Petri (father of the pianist Egon Petri) and four conductors. Mahler's superior in age and seniority was Arthur Nikisch, who to Mahler seemed cold both as a conductor and personally. Inevitably the local critics considered Nikisch to be the yardstick against which any new conductor was to be measured, and in view of his different character and artistic temperament Mahler was found wanting. The main attraction of the season was to be the production of the first *Ring* cycle in Leipzig, and on hearing that Nikisch was to conduct, Mahler handed in his resignation. It was not accepted. Fate intervened after the production of *Das Rheingold* in January 1887, when Nikisch fell ill, leaving Mahler to conduct *Die Walküre* and *Siegfried*. With his outstanding production of the latter in May, Mahler convincingly established among critics and public his genius as an interpretative artist.

Weber's operas formed an important part of the repertory of the Stadttheater and Mahler conducted a number of them in December 1886. At about this time he met Carl von Weber, the composer's grandson, who possessed his grandfather's sketches for a comic opera entitled *Die drei Pintos* which the family had sent to Meyerbeer and Franz Lachner for completion. Neither had been willing to undertake the task, particularly since Weber's musical shorthand was often very difficult to decipher, and Mahler's initial reaction on seeing the manuscript was similar. Greater familiarity awoke his enthusiasm, however, and having mastered the short-hand and aroused Staegemann's interest in the project, he spent summer and early autumn 1887 adapting lesser-known pieces by Weber and composing passages based on Weber's themes to complete the work. Although not original, *Die drei Pintos* was important to Mahler, for its successful première in January 1888 and subsequent performances throughout Germany made him famous and provided a useful source of income. Moreover, the orchestral score, the first by Mahler to be published (1888), is fascinating for the insight it offers into his early, but already characteristic, orchestral style.

The work on the opera inevitably involved close contact with the Weber family, and Mahler fell in love with Carl von Weber's wife. This hopeless affair inspired in Mahler a period of intense creative activity, which saw the completion of two works, a symphonic poem – the eventual First Symphony – and the first movement of a symphony in C minor entitled *Todtenfeier*. It was not until Mahler's Hamburg period (1893–4) that the *Todtenfeier* was revised and the rest of the Second Symphony was composed. The Symphonic Poem in D similarly underwent two revisions – in 1893, and around 1896 when the second movement entitled 'Blumine' was deleted – before becoming the First Symphony. A further consequence of Mahler's acquaintance with the Weber family was his discovery in 1887 of the musical potential of *Des Knaben Wunderhorn*, an idiosyncratic collection of texts chosen and adapted by Achim von Arnim and Clemens Brentano to reflect their conception of folk poetry. It was this work that provided Mahler with all but one of his song texts during the next 14 years; the first fruit of this literary influence was the composition between 1887 and 1890 of the nine settings forming the second and third volumes of the *Lieder und Gesänge*, which were published complete in 1892.

In the early months of 1888 Mahler met two important musicians, Tchaikovsky, who did not recognize Mahler's genius as a conductor until they met again in Hamburg, and Richard Strauss. Although quite different in temperament, Strauss and Mahler recognized genius in each other – without, in Strauss's case at least, fully appreciating its precise nature – and their friendship lasted until Mahler's death.

In May that year, following a violent argument with a stage manager during a rehearsal for Spontini's *Fernand Cortez* at Leipzig, Mahler again offered his resignation, and this time it was accepted. The suddenness of the affair left Mahler unprepared and without a suitable post until August, so he was able to devote himself to uninterrupted work on *Todtenfeier*. The summer post was an engagement to prepare the Prague première of *Die drei Pintos* and a production of Cornelius's *Der Barbier von Bagdad*, but the renewed collaboration with Neumann had lasted less than a month when, after an artistic disagreement, Mahler was dismissed. Although once more without an engagement, he was no longer a 'fahrender Gesell' but a 'Meister', and within a few weeks had concluded negotiations for a far more important and attractive appointment at Budapest.

CHAPTER FOUR

Budapest and Hamburg, 1888–97

Like so many other occasions during his conducting
career, Mahler found the institution for which he was
engaged in a depressed state. The Budapest Royal Opera,
one of the most modern and best equipped theatres in
central Europe, was running at a loss, with a limited
repertory and a badly disciplined company which had to
be strengthened by guest artists. In addition, there were
increasingly vociferous demands for the establishment
of a Hungarian national opera, and it was with these in
mind that Mahler set to work: performances were to be
in Hungarian, native singers were to be engaged when
possible, an ensemble was to be developed and guest
appearances reduced. Although appointed in September,
he delayed his appearance as a conductor until the
première of *Das Rheingold* in a new Hungarian translation in the following January, but in the meantime he
supervised several new productions. The triumph of *Das
Rheingold* and *Die Walküre*, performed on consecutive
evenings, vindicated Mahler's efforts, and the rest of his
first season was a great success.

The first year in Budapest was overshadowed by the
death of Bernhard Mahler in February 1889 and the
subsequent illness of Mahler's mother and his sister
Leopoldine, who both died the following autumn.
Already depressed, Mahler had to face problems at the
opera, the most serious of which was his inability to find

enough talented Hungarian singers. Coupled with this was the public's resistance to the mainly German repertory he had established, which forced the postponement of the new productions of *Siegfried* and *Götterdämmerung* in favour of light opera and ballet. The conservative public was also ill-prepared for the première of Mahler's own Symphonic Poem (First Symphony), on 20 November 1889, and during the performance the audience became increasingly hostile towards the innovatory work.

In search of new works for the Budapest opera, Mahler travelled to Italy in summer 1890; he chose Franchetti's *Asrael* and Mascagni's *Cavalleria rusticana*, and the subsequent production of the latter under Mahler's direction (26 December 1890) was the first outside Italy. On his return to Hungary he began preparations for a new *Don Giovanni*, and it was this production that aroused Brahms's admiration; a few years later Brahms's recommendation, together with that of Hanslick, was an important factor in favour of Mahler's appointment at Vienna. Despite this success, his situation at Budapest became untenable because of the imminent appointment of Count Géza Zichy as Intendant. Zichy was a nationalist and had indicated that he, rather than Mahler, would have artistic control of the house. Mahler, anticipating the inevitable disagreement, contacted Pollini in Hamburg, was offered the post of first conductor there, and accepted it.

Mahler was only partly successful at Budapest, for although under his direction the opera made a profit and the performing standards and repertory improved immeasurably, much remained to be done. Moreover, his superhuman efforts as an administrator and interpretative artist had been made at the expense of composition.

Some of the early *Wunderhorn* settings from *Lieder und Gesänge* had been completed at Budapest, as had the scoring of the First Symphony, but this was a small output.

Bernhard Pollini, who was the director of the Stadttheater at Hamburg, was a remarkable impresario with a talent for gathering great artists about him, and as a result his company was of international standard. Mahler's position as first conductor brought with it no control over the staging of opera, an aspect that Pollini neglected, and little independence in artistic matters. With his early performances of Wagner, Mahler convinced the public, critics and Pollini of his genius as a virtuoso conductor; on 18 May 1891 he conducted *Tristan*, one of his most celebrated interpretations, for the first time, though by then he was undertaking the heaviest schedule of his life – up to 19 operas a month – which left little time for proper rehearsal. After four years of almost continual strain he spent summer 1891 not composing, but as a tourist, visiting Bayreuth and Scandinavia. In the autumn Mahler returned to a rigorous schedule which included new productions of Tchaikovsky's *Eugene Onegin* and Bruneau's *Le rêve*, which won the admiration of their composers. Mahler was conducting an increasing number of concerts, and following Bülow's illness and death he took over the direction of the subscription concerts in Hamburg (1894). Bülow had been impressed by Mahler's productions of Wagner and was a sincere admirer of his genius as a conductor, but he failed to understand his work as a composer.

At the beginning of summer 1892 Mahler was in London at the invitation of Sir Augustus Harris to conduct a season of German opera. The casts included some of the best German singers of the day, the orchestra was specially assembled in London and the performances were given at Covent Garden and Drury Lane before enthusiastic audiences, but Mahler nevertheless begrudged the time spent on them that might have been devoted to composition, and as a result did not accept Harris's subsequent invitations.

Despite the stifling artistic atmosphere and the heavy work at Hamburg, Mahler returned to composition (orchestral *Wunderhorn* settings, Second Symphony) and as no other post was available he signed a new five-year contract in 1894. At that time there was a possibility that Strauss would also be engaged at Hamburg and relations between the two composers were particularly close. Mahler's attempts to stage *Guntram* failed, but Strauss's influence brought about performances of Mahler's Symphonic Poem at Weimar (3 June 1894) and the first three movements of the Second Symphony in Berlin (4 March 1895). The Symphonic Poem had already been given a second performance in Hamburg the previous year, and both there and at Weimar it bore the additional title 'Titan', a reference to Jean Paul's novel, the spirit of which pervades the work.

Work on the Second Symphony, finished in 1894, was succeeded in the following two summers by the composition of the massive Third Symphony, the pattern of Mahler's creative life having been established: composition of short scores in the summer months, revision and orchestration of them during the theatre season. In 1893 Mahler had discovered a summer retreat at Steinbach in the Salzkammergut, and it was there that the two works were completed. The première of the complete Second Symphony in Berlin on 13 December 1895 was Mahler's first great public success as a composer, and the symphony remained one of his most popular large-scale works. The Third Symphony, however, had to wait until 1902, when it was given its first complete performance at Krefeld.

Tragedy again struck Mahler's family in 1895, when his youngest brother Otto, a young man of genuine musical ability, committed suicide in Vienna. Since his parents' deaths, Mahler had supported his two remaining sisters, and had encouraged Otto to take up a musical career. Mahler's financial responsibility for his sisters Emma and Justine continued until their marriages in 1898 and 1902 to two brothers, Eduard and Arnold Rosé, both of whom were accomplished string players (Arnold was leader of the Vienna PO for 50 years).

Besides the existing repertory Mahler supervised a series of new productions at Hamburg including *Falstaff* (1894), Mascagni's *L'amico Fritz* (1893), *Hänsel und Gretel* (1894) and four of Smetana's operas. In 1895 a new soprano, Anna von Mildenburg, was engaged; Mahler, perceiving her talent, devoted many hours to training her, and she became one of the great Wagner singers of the day. In spite of this successful artistic collaboration he once more began to feel restless and turned his attention to the Vienna Hofoper. He was by this time a conductor of international stature (he undertook his first tour, to Moscow, Munich and Budapest, early in 1897) and a composer whose works were being performed with increasing frequency, but his intractable nature and Jewish origins were obstacles to any appointment at Vienna. In order to remove one he accepted baptism as a Catholic (23 February 1897) and after months of negotiations was appointed Kapellmeister at Vienna (April 1897).

CHAPTER FIVE

Vienna, 1897–1907

Although in May 1897 the Vienna Hofoper was managing to maintain its widespread reputation as one of the great opera houses of the world, it was stagnating under an ailing director, Wilhelm Jahn. Of the other three resident conductors, only Hans Richter, who left for England in 1900, was of international stature. With masterly performances of *Lohengrin* (his Viennese début, 10 May) and *Die Zauberflöte* (29 May), Mahler immediately established himself as an interpreter of Wagner and Mozart. But ill-health prevented his conducting more frequently during the rest of the season; the summer was spent convalescing and preparing for the forthcoming season (he was appointed director on 8 September 1897) which was to include an uncut *Ring* cycle for the first time in many years. Wolf, impressed by Mahler's interpretations, visited him in the hope that he would produce *Der Corregidor*, but an argument broke out over the artistic merits of Rubinstein's *The Demon* (shortly afterwards Wolf was committed to an asylum). It was only in 1904 that Mahler finally mounted Wolf's opera, and then in a revised version: despite Mahler's efforts (later performances were given using the original text) the work failed to enter the repertory.

Late in 1897 Guido Adler, an old friend of Mahler's, arranged a grant which paid for the publication of the scores of the First and Third Symphonies, together with the orchestral parts of the Second, an act of friendship which, in view of the increased interest shown in the works, was timely indeed. Adler's influence also led to Mahler's appointment to the editorial board of the Denkmäler der Tonkunst in Österreich, though Mahler found the lengthy editorial discussions tedious, and the published music mediocre.

When in 1898 Richter resigned from the direction of the Philharmonic Concerts Mahler was invited to replace him. As a concert conductor, Mahler was already controversial, and continued to be so in Vienna. An immediate effect of his engagement was to increase the size of audiences dramatically and this, together with a successful tour to Paris in summer 1900, boded well for the future, but relations between Mahler and the players soon began to deteriorate because of his authoritarian attitude and unconventional views. After his serious illness in 1901, during which the concerts were directed by comparatively mediocre conductors to the obvious pleasure of both critics and players, Mahler resigned.

Mahler spent his 1899 holiday at Aussee in the Styrian Salzkammergut. At first it seemed that this summer would be no more productive than the previous one, during which only two more *Wunderhorn* settings were sketched; but after composing *Revelge* and with only ten days remaining, Mahler found the Fourth Symphony suddenly beginning to take shape. In the early weeks of the new season he searched for a new summer home where he could compose in peace, and eventually chose a site at Maiernigg on the Wörthersee in Carinthia. There, in the following year, work on the Fourth Symphony was continued, and by 1901 the villa where

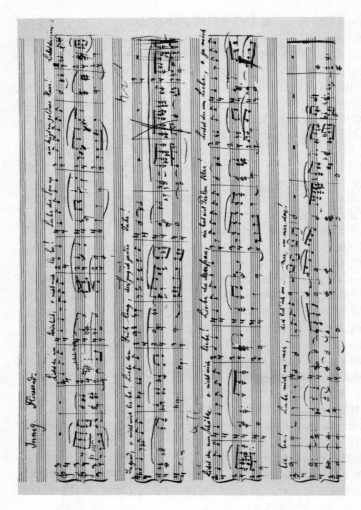

7. Autograph MS of Mahler's song 'Liebst du um Schönheit', composed August 1902

he was to compose all his significant works for the next five years was completed. In the remarkably fruitful summer of 1901, four of the *Rückert-Lieder*, three *Kindertotenlieder*, *Der Tamboursg'sell* and two movements of the Fifth Symphony were drafted.

In autumn 1901 Mahler met and fell in love with the daughter of the Austrian landscape painter Anton Schindler. Alma Schindler (1879–1964) was an intelligent young woman who belonged to a talented artistic circle and was studying composition with Alexander von Zemlinsky. The marriage, which took place on 9 March 1902, caused much surprise among the couple's friends, and their life together was not untroubled, for Mahler required it to be arranged around his own creative work, and insisted that Alma should give up composition. Eventually the strains imposed by the attitudes of both partners and the difference in their ages began to affect their relationship, and Mahler consulted Freud in summer 1910. Freud was struck by Mahler's immediate understanding of the principles of psychoanalysis (which is not surprising considering the composer's philosophical interests and artistic outlook), while Mahler gained an insight not only into his immediate marital problems, but also into his own creative personality. He rediscovered the depth of his love for Alma, as is shown by the touching messages addressed to her in the manuscript of the unfinished Tenth Symphony.

The repertory of the Hofoper during Mahler's directorship was not remarkable for the number of important new works included, though in 1902 Richard Strauss's *Feuersnot* was produced, and this was followed by Charpentier's *Louise* (1904) and Pfitzner's *Die Rose vom*

Liebesgarten (Mahler failed in his attempt to stage *Salome* because of the Viennese censor). Mahler's main achievement lay in the re-creation of established masterpieces in integrated productions. In this respect his marriage was of decisive importance, for it was through Alma that Mahler came into close contact with several members of the 'Sezession' movement. Not only did he take part in the Max Klinger exhibition of 1902, but also invited one member of the movement, Alfred Roller, to design a new production of *Tristan und Isolde* (1903). This marked the beginning of the most brilliant period of Mahler's reign in Vienna, for Roller, appointed chief stage designer in 1903, combined great artistic sensitivity and inventiveness with a profound understanding of, and sympathy for, Mahler's conception of operatic production. Together Mahler and Roller created a series of memorable productions, *Fidelio* (1904), *Don Giovanni* (1905), *Die Entführung aus dem Serail* (1906), *Le nozze di Figaro* (1906), *Die Zauberflöte* (1906) and *Iphigénie en Aulide* (1907).

Mahler had no aspirations to be a teacher or the leader of a group. Nevertheless during his years in Vienna he was surrounded by an increasingly large circle of radical young composers (including Schoenberg, Berg, Webern and Zemlinsky) who saw him not merely as a great artist but also as their supporter in the Viennese musical establishment. Although not always in agreement with their views, Mahler expressed his support through his work for the Vereinigung schaffender Tonkünstler (1904–5), and he gave material assistance to Zemlinsky and Schoenberg.

As at Hamburg, Mahler's creative life benefited immeasurably from security, which in Vienna he found in

GESELLSCHAFT DER MUSIKFREUNDE
IN WIEN

Sonntag, den 24. November 1907, mittags halb 1 Uhr
===== im großen Musikvereins-Saale =====

I. AUSSERORDENTL. GESELLSCHAFTS-KONZERT.

o o o o o

Zur Aufführung gelangt:

GUSTAV MAHLER =

ZWEITE ·SINFONIE (C-MOLL)

== für Soli, Chor, Orchester und Orgel. ==

1. Satz: ALLEGRO MAESTOSO. (Mit durchaus ernstem und feierlichem Ausdruck.)
2. Satz: ANDANTE CON MOTO.
3. Satz: SCHERZO. (In ruhig fließender Bewegung.)
4. Satz: „URLICHT" aus: „Des Knaben Wunderhorn".
5. Satz: FINALE.

MITWIRKENDE: ========

Frau ELISE ELIZZA, k. k. Hof-Opernsängerin.
Fräulein GERTRUD FÖRSTEL, k. k. Hof-Opernsängerin.
Fräulein HERMINE KITTEL, k. k. Hof-Opernsängerin.
Fräulein BELLA PAALEN, k. k. Hof-Opernsängerin.
Herr RUDOLF DITTRICH, k. k. Hoforganist.
Der SINGVEREIN DER GESELLSCHAFT DER MUSIK-
FREUNDE.
Das K. K. HOF-OPERNORCHESTER.

DIRIGENT: DER KOMPONIST.

Preis dieses Programmes 20 Heller.

Buchdruckerei: Wien, I., Dorotheergasse 7.

*8. Programme of Mahler's farewell concert on 24 November
1907*

105

his marriage, and during summer vacations at Maiernigg he wrote four new symphonies and a number of songs. The Fifth Symphony, completed in 1902, was succeeded the following year by two movements of the Sixth. The draft of this work, which incorporates musical 'portraits' of Alma (first movement, second subject, bars 77ff) and children playing (Scherzo), was completed in 1904, a summer of unparalleled creative activity which also saw the drafting of the two remaining *Kindertotenlieder* and the two 'Nachtmusik' movements of the Seventh Symphony. The completion of the latter proved problematic, however, and it was only at the end of the 1905 holiday, after a long search for inspiration, that the remaining three movements were sketched. It was probably a memory of that search which drew Mahler in 1906 to two texts which became the basis of the Eighth Symphony: the medieval hymn *Veni creator spiritus*, and the closing scene from part 2 of Goethe's *Faust*, the first an invocation of and the second a celebration of the creative spirit. For all its complexity, this work, which Mahler at the time of its composition described as 'the greatest work I have yet composed, . . . so different in contrast and form that I cannot even write about it', was sketched in only eight weeks.

The campaign against Mahler in his position as director of the Hofoper, which was being led by the anti-semitic press in Vienna, gradually gained momentum as he devoted more time to the propagation of his own music, and it was over this that matters finally came to a head. Mahler was able to refute claims that his absences on leave resulted in diminished box office receipts, but both he and the administration had decided on his departure, though this did not officially happen until December 1907. In June of that year Heinrich Conried, the manager of the Metropolitan Opera, engaged Mahler to conduct there from January 1908.

Soon after their arrival at Maiernigg in summer 1907, the Mahlers' elder daughter fell ill. Mahler was devoted to both the children, Maria (*b* 1902) and Anna (*b* 1904), particularly Maria, and her death from scarlet fever a few weeks later left him distraught with grief. The third and final blow following his resignation and bereavement – and as the Sixth Symphony had prophesied it eventually felled its victim – was the discovery that Mahler suffered from a heart condition and would have to give up his favourite pastimes, walking, cycling and swimming.

In the following October and November Mahler conducted his last opera at Vienna, *Fidelio*, and, at a farewell concert, his own Second Symphony. His directorship at Vienna was a landmark in the cultural life of the period, the collaboration with Roller a milestone in the history of the stage presentation of opera and his reign at the Hofoper one of the most glorious in its history.

CHAPTER SIX

New York and Europe, 1907–11

The traumatic events of the year 1907 were responsible for effecting a drastic change in Mahler's attitude to life and art and this was reflected in his work at the Metropolitan Opera. The company he found there was the opposite of the type he had constantly sought to create in the past, consisting as it did of great operatic stars rather than an integrated ensemble; the emphasis lay on the vocal performances. Nevertheless Mahler not only made little attempt to alter this situation but even agreed to extensive cuts in Wagner. His first performance in New York was *Tristan und Isolde* (1 January 1908), which was followed by *Don Giovanni*, *Die Walküre*, *Siegfried* and *Fidelio* (the last using Roller's original set designs).

The spring and summer of 1908 were spent in Europe conducting concerts, including the première of the Seventh Symphony at Prague, and composing at a new summer retreat at Toblach (now Dobbiaco) in the Dolomites. Whether in spite of or because of the tragedy, summer 1907 had not remained completely fruitless, and Mahler had begun to sketch ideas for a new work. Shortly before, he had been given a volume of Chinese poems in German versions by Hans Bethge, *Die chinesische Flöte*, and it was from this collection that he chose the seven texts for *Das Lied von der Erde*. As the work proceeded the piece grew into a large-scale symphony for tenor, contralto (or baritone) and orchestra, but Mahler, citing the examples of Beethoven, Schubert and Bruckner, refused to number it among his symphonies, hoping thereby to cheat fate.

Soon after Mahler's arrival in New York, Conried resigned as manager and Giulio Gatti-Casazza was invited to replace him. At first Mahler viewed the changes unfavourably: eventually he decided to return the following year, by when the new manager, together with Toscanini, had arrived from La Scala. During the 1908–9 season Mahler had conducted two new productions, *Figaro* and *The Bartered Bride*, as well as *Tristan*, and also three concerts by the New York SO. Shortly after the concerts he accepted the conductorship of the revitalized New York PO, and it was probably this opportunity that persuaded him to cancel his contract with the Metropolitan. His last appearance there in 1910, conducting a new production of *The Queen of Spades*, was as a guest artist. A further indication of the change in Mahler's attitudes is seen in his American concert programmes which, unlike his earlier European ones, contained a large number of new works. He admitted that his main pleasure was in rehearsing works unfamiliar to him, and his late compositions show traces of the influence of Debussy's orchestral music, some of which (*Rondes de printemps*, *Ibéria*) he conducted in the 1910–11 season. This interest in new music also led to a rekindled fascination for the music of the past, some of which he conducted in a series of historical concerts, and ultimately to his arrangement of four movements by Bach into an orchestral suite.

The leisurely summer of 1909 saw the completion of the Ninth Symphony, but in the following year Mahler's

9. Gustav Mahler, 1907

110

stay in Europe was extraordinarily busy. When not conducting concerts (Paris, Rome, Cologne) or sketching the Tenth Symphony, he was engaged in the preparations for the première of the Eighth at Munich. The work, performed on 12 and 13 September before a distinguished audience, was tumultuously acclaimed, confirming Mahler's own suspicions voiced in 1909: 'this work always makes the typical, strong appeal. It would be an odd thing if my most important work should be the most easily understood'.

Mahler returned to New York for the winter season and conducted a strenuous concert series with the New York PO, which included two performances of his Fourth Symphony in January 1911. The following month he fell ill and was found to be suffering from a bacterial infection from which there was little hope of recovery. He therefore decided to return to Vienna, stopping first in Paris where the diagnosis and prognosis were confirmed. He arrived in Vienna on 12 May and died there on 18 May 1911, a few weeks before his 51st birthday.

There can be little but speculation about the actual character of Mahler's conducting, though no doubt at all about the impact it made on his contemporaries. One thing is certain, that his performances of other men's music was re-creative rather than interpretative. That is to say, he approached the works he conducted as if he had composed them himself and laboured to achieve the precision, the clarity and the vivid characterization that he insisted on in performances of his own music; this is evident in the elaborate notation of his scores which attempts to find a sign for even the subtlest of nuances. If this primarily re-creative spirit is understood, much else becomes clear, in particular his activities as editor or arranger. His notorious retouching of Beethoven's orchestration for his performances or, a very different example, his devising of a lengthy new recitative for Mozart's *Figaro* (based on the omitted trial scene from Beaumarchais), should not be regarded as meddling or evidence of his belief in the technological or artistic superiority of the 19th century, but rather as modifications made in the image of his own compositional practice and convictions, in which a passion for clarity, whether of sound or operatic plot, was paramount.

CHAPTER SEVEN

'Das klagende Lied' to Symphony no.1

Gustav Mahler himself quite rightly considered *Das klagende Lied* to be the first work in which he found his own voice as a composer. This precocious cantata of 1880 displays, for all its evident immaturities, an extraordinary number of conspicuously Mahlerian mannerisms. The discovery of his musical identity coincided with the writing of the cantata: hence the comparatively eclectic style and less cogently organized form of the original first part compared with the much more personal and shapely second and third parts (which became the two parts of the revised and published version of 1899). As a result of the omission of the first part, the cantata's originally concentric tonal scheme, A minor–A minor, was lost; thus the 'progressive' scheme that remains, C minor–A minor, was not consciously planned but a consequence of the revision. Since the rediscovery of the first part, views have been expressed about the desirability of returning to the composer's original conception; but Mahler's revision showed his excellent critical judgment, and a precise assessment of where he began to speak his own language consistently.

The style of *Das klagende Lied*, even at its most personal, shows its origins in the music to which Mahler was responsive, in particular the operas of Weber and Wagner. The cantata derives from some of the principal

113

sources of German Romanticism in the late 19th century: magic, a revived medievalism and nature. But to all this the youthful composer gave a new twist by setting the catastrophe of the final scene of the work against an offstage wind band, which pursues its festive music despite the calamitous dramatic context. This was a highly original climax, and in a single knot of sound tied together, as early as 1880, many of the threads that were repeatedly to appear in Mahler's music until his death. These include the combining of onstage–offstage orchestras and exploiting the particular acoustic relationship involved in this device; the association of high tragedy and the mundane (the wind timbre of the offstage band), the latter offering a simultaneous ironic commentary on the former; and the operatic build of the dramatic dénouement.

Four other aspects of the cantata also demand mention. First, even in the original version the handling of the orchestra – an exuberantly large one for the time – shows unusual flair and imagination for a young composer. Second, there is a prominent kind of vocal melody which anticipates the directness and intensity of utterance of the *Lieder eines fahrenden Gesellen* (1884–5). Some commentators have seen the influence of Bohemian folksong on the melodies of *Das klagende Lied*; certainly the vocal shapes in ex.1 are distinct from

Ex.1 *Das klagende Lied*

114

anything that Mahler had written earlier for voice in the first volume of the *Lieder und Gesänge*. The latter includes songs relatively traditional in format, despite their merits and originalities, among them the first published appearance of Mahler's progressive tonality, in *Erinnerung*, which begins in G minor and ends in A minor. Even the most folklike of the songs, *Hans und Grethe*, which was to be the source of a whole movement in the First Symphony, did not break such new ground as ex.1: the latter, shaped by a crucial dramatic moment, is free from the rounded, symmetrical periodization of the lied. Folk ballad, a form of dramatized story-telling, rather than folksong, is an appropriate description of its style. Third, there is the complex, synthesizing form of the cantata itself, which exploits a variety of musical forms and resources within a dramatic frame. Fourth, there is already a clearly defined dramatic–symbolic use of tonality: for instance F♯ minor may be understood as the 'murder' key (the slaying of the younger brother by the elder). The cantata proved to be a seedbed of inventive possibilities which found their ultimate fulfilment in Part 2 of the Eighth Symphony, Mahler's most ambitious attempt to use the techniques and styles of dramatic cantata in the service of a symphony that was itself the most thoroughgoing of his composite works.

In the intervening years, there were important 'hybrid' works such as the Second and Third Symphonies, which incorporated forms and resources traditionally kept apart (song and symphony), but in general none of the music is as 'pluralist' in conception as the early cantata. It was rather the case that in succeeding works, the forms of which were ever more

10. Gustav Mahler, 1884

narrowly defined, possibilities were explored which had been first adumbrated in the cantata.

His next work, the *Lieder eines fahrenden Gesellen*, which, because of the beauty of its formal design, should properly be regarded as the masterpiece of his early years, is an orchestral song cycle, and, like the cantata, on texts by Mahler himself. There is little doubt tnat the cycle was envisaged at its earliest stage with orchestral accompaniment, though first drafted (as was Mahler's custom with his later orchestral songs) for voice and piano and actually orchestrated much later. The features of the cycle may be summed up as follows: the development from the cantata of the narrative idea, an unfolding drama, but this time applied to a sequence of four songs; the musical representation of this 'dramatic' idea, which implies a destination wholly different from the point of departure (a notion opposed to classical symmetry based on the reprise), through an evolutionary tonal scheme which 'travels' along with the changing fortunes of the travelling hero (the first song goes from D minor to G minor, the second from D major to F♯ major, the third from D minor to E♭ minor, and the fourth from E minor to F minor); the exploration of textures and developments more symphonic in their density and elaboration than normally associated with the lyric song (here Mahler was doubtless influenced by the large-scale thinking in Wagnerian music drama); and finally, the unmistakable identification of the composer with his *fahrender Gesell*. No longer relating a story, as in *Das klagende Lied*, Mahler appears as the hero of his own work. The Artist as Hero is a role that he continued to play to the very end of his creative life, though always in a phenomenal variety of guises (not disguises).

If the *Lieder eines fahrenden Gesellen* were not orchestrated soon after being drafted, it was probably because Mahler had started work (possibly as early as 1884, the year in which the cycle was completed) on a large orchestral composition. This was to become the First Symphony but started life and was first performed as a symphonic poem in two parts and in five movements; 'Blumine', the original second movement (Andante), was later discarded. Particularly in its first version, the work shows that Mahler, like his aspiring contemporary Richard Strauss, was dominated by the idea of programme music, though he was also acutely conscious of the necessity to prove himself a credible symphonist. During this period of composition, up to about 1901, it was out of the friction between symphony and symphonic poem that some of Mahler's most interesting experiments arose; the uncertainties and ambiguities plaguing him at this time proved creatively valuable. Like any ambitious young composer of his day Mahler was following the new trend, but he brought to his own use of symphonic poem a highly personal approach. For example, in the First and Second Symphonies and even in the Fourth he maintained and further developed the narrative principle of *Das klagende Lied* and the *Lieder eines fahrenden Gesellen*. Also, and most important, he took over actual musical materials from his own earlier works: thus he made a substantial and dramatically apt quotation from the last song of the *Gesellen* cycle in the funeral march slow movement of the First Symphony, used an early song in the Scherzo and built the work's ample first movement out of the cycle's second song, *Ging heut' morgens übers Feld*; this song is the most complex and polyphonic in texture, thus offering the most possibilities for development in a more expansive form. This unique incorporation of song cycle into symphony is revealing of Mahler's working methods, and it also underlines the need to consider the body of his work as a whole. It would be exaggerating to suggest that there are no distinctions to be made between one type of work and another, but it is the mixture of categories that is a particular feature of the works up to the turn of the century.

CHAPTER EIGHT

'Wunderhorn' works

The interrelationship of song cycle and symphony is seen in Mahler's *Wunderhorn* songs, settings of poems from the Arnim and Brentano anthology of 1805–8. He was to use one of the earlier voice and piano settings (of 1887–90), *Ablösung im Sommer*, as the basis for the Scherzo of the Third Symphony; but it was the later orchestral settings (from the 1890s) that were to be of such creative significance. Mahler gave a new twist to his *Wunderhorn* settings by lifting them out of the realm of fairy tale and neo-medievalism and by reinterpreting them through his personal experience. Thus the archaic images of the poetry, embodied in intensely personal music, often came to be symbolic in Mahler's aesthetic of Man's (or the Artist's) experience of the world. Despite the clear significance of the texts, the vividness of Mahler's rendering of the poems derives from the pungency and clarity of his orchestration, and the memorability and characterizing power of his vocal melodies. The orchestral songs met with a mixed reception at the time of their first performances. A minority readily appreciated their originality. The majority was disturbed by what seemed a discrepancy between the 'primitive' folksong materials and Mahler's highly elaborate, often ironic and intense treatment of them. But this adverse criticism was only a reaction against his defiance of long-established conventions for folksong settings.

'Wunderhorn' works

The *Wunderhorn* songs are important in themselves as a set of orchestral songs, for which there were few precedents outside the musical theatre: Berlioz was undoubtedly one of Mahler's models, and an influence throughout this period. Furthermore, they functioned as a storehouse of invention, symbol and image in parallel relation to the symphonies Mahler wrote during his 'Wunderhorn' years. He was engaged in both genres during this time, and where song could clarify an important moment in the structure of a symphony he would without hesitation turn to his songs, or in certain cases write new ones, in order to complete or make coherent his dramatic schemes. Examples are *Urlicht* in the Second Symphony, or *Es sungen drei Engel* and *O Mensch! Gib Acht!* in the Third. Song could also form the basis of a large-scale instrumental movement, as with the *Fischpredigt* Scherzo in the Second and the *Kuckuck* Scherzo in the Third Symphony; or it could function as the poetic and musical summit of an entire symphony, as with *Das himmlische Leben* in the Fourth. At this period Mahler's creative process can be seen as a kind of assembly line: music was composed of a specific category (instrumental movements, songs or song cycle), and then movements of various kinds were chosen to fill out the dramatic outline, tried out in various sequences, added to or contracted and finally shaped into cogent large structures. It was an evolutionary process, which because of its fluid nature until the final stage sometimes meant that what was excluded from one work formed the basis of or was included in the next. The interrelationship of the Third and Fourth Symphonies provides a clear instance of this practice. That part of the story which did not get told in the Third was explored in the Fourth; and the Second Symphony (and probably the First, too, if one knew more about the chronology of the earliest stages of its composition) provides another example of Mahler's assembly-line procedures. Thus while Mahler's 'Wunderhorn' symphonies have their roots in Beethoven, Berlioz and Liszt, they were also significantly influenced by the *Wunderhorn* songs, which introduced into the various dramatic schemes a particular range of imagery very different from, for instance, the kind of programmes that Strauss was using.

After finishing the Symphonic Poem (First Symphony) Mahler immediately embarked on completing, on the grandest scale, the first movement of a symphony in C minor which was plainly intended to establish his symphonic credentials. The opening movement of the Second Symphony is one of the most imposing of Mahler's sonata structures, unorthodox in its tonal organization, but unambiguously and even 'classically' articulated. Mahler became stuck after the first movement was written and was only able to complete the work after a long interval and the devising of a programme. While the sequence of the middle movements and the gigantic finale with its choral culmination extend the idea of continuous narrative that was the frame for the five movements of the Symphonic Poem, the funeral march is a structure more explicitly and traditionally symphonic than might have been expected after the free, untrammelled form of the immediately preceding work. This formal trend was not sustained across the whole arch of the Second Symphony. Nor was it continued in the Third, the most idiosyncratic of Mahler's symphonies, whose dramatic scheme evolved gradually and with recourse to song (Nietzsche text) and chorus (*Wunderhorn* text) in the fourth and fifth movements. To approach the huge first movement with preconceived ideas of symphonic sonata form is to court analytical disaster and aesthetic disappointment; it makes sense only in terms of its dramatic programme, the symbolic conquest of winter by summer, which was detailed in the original manuscript full score, though references to it were expunged from the published version. That Mahler had ambiguous feelings about programmes (he often expressed his opposition to the concept and all the misunderstanding it entailed and eventually gave up using it altogether) in no way diminishes the importance of the programmatic idea for him during the 'Wunderhorn' years.

With the Fourth Symphony Mahler returned to symphonic tradition, in a first movement of incomparable wit and subtlety that is not only a highly original approach to sonata form but also a fascinating commentary, in purely musical terms, on the sonata idea. An example is the sophisticated treatment of the moment of recapitulation, which indeed recapitulates everything but chronologically displaces the expected sequence (ex.2). It is a typical Mahlerian paradox that in the work which contains perhaps his most complex and meticulously organized sonata movement, he also developed to its highest pitch, and in its most sophisticated guise, the idea of the programme. In this case, the poetic idea is the progress from experience (the first movement) to innocence (the *Wunderhorn* song finale, *Das himmlische Leben*). The progression is represented in a sequence of forms and textures which gradually diminishes in complexity until the childlike vision of Paradise is attained in the 'heavenly' key of E major, a tonal destination

123

Ex.2 Symphony no.4, 1st movt
(a)

already foreseen in the opening B minor bars of the first movement, before the movement's – and the symphony's – central tonality, G major, is unambiguously asserted. This is the symphonic work that closes the 'Wunderhorn' period, and in which Mahler finally reconciled two preoccupations sometimes at odds with one another, the symphonic tradition and the very different requirements of a developing drama.

In *Das klagende Lied*, the offstage 'festive' band comments ironically on high tragedy. In the First Symphony, the Funeral March presents a parody, based

(b)

on a grotesque distortion of a children's round (*Bruder Martin*); moreover, the irony is sustained to the bitter end of the movement, except for the middle part of the march based on the last of the *Gesellen* songs (itself a vocal quasi-funeral march, a peculiarly Mahlerian invention which also crops up among the *Wunderhorn* songs). Irony was not of course new to music: what was new in Mahler's use of the ironic mode was its comprehensiveness. Parody, irony, satire: these were all elements that had their musical foundation in the works from Mahler's 'Wunderhorn' years and became embedded permanently in his musical style. Moreover,

125

11. Autograph
sketches for the
first movement of
Mahler's Fourth
Symphony

these were components almost always inextricably mixed up with his singular use of popular musical materials, to which the ironic method and interest gave a new twist. A locus classicus is the 'vulgar', rowdy brass-band 'Resurrection' march from the Second Symphony, where the dramatic context – a march of the dead towards eternity – charges with almost elemental power the seemingly mundane musical formulae that Mahler so energetically manipulated with a highly ingenious use of Lisztian thematic transformation (ex.3). But it would be wrong to assume that Mahler's ironic mode and his use of popular musical materials always went hand in hand. For example, in the audacious first movement of the Third Symphony he used all manner of popular invention – swinging march tunes, military signals and fanfares, and vigorous march rhythms – to embody his dramatic concept of summer marching in and sweeping winter away. There is no irony here; the popular materials are used in an almost Ivesian, collage-like fashion, with orchestral mastery and compositional virtuosity serving a symphonic movement built on the grandest scale.

Another important feature of these years was the development of Mahler's orchestration. From the start of his creative life he showed exceptional gifts in this sphere, but he swiftly progressed from the lavish deployment of resources in the early cantata to the famous clarity, refinement and soloistic scoring of, for example, the Fourth Symphony, a trend that was itself embodied in a decisive move away from strings, as the basis of his orchestral sonority, to wind. The development of this highly personal orchestral sound meant a growing emphasis on selectivity, the precise use of his resources to

Ex.3 Symphony no.2, last movt

meet the precise needs of the moment. It is correct to refer to Mahler's orchestral practice as economical; on the other hand, so fine and inventive was his ear that he usually needed very large orchestras from which his characteristic wealth of constituent ensembles might be drawn. In his symphonies he may have modified, but never in principle abandoned, the large orchestra as a basic unit; and the fact that his scoring in successive works grew more transparent and refined is directly

128

related to the large number of slender and economical textures (not to mention the wealth of subtle instrumental timbres) offered by the sheer abundance of his resources. The clarified final version of a full score often involves a larger orchestra than that used in its denser-textured first version. It is impossible to separate Mahler's orchestration from his actual methods of composing. For example, in the opening of the Funeral March in the First Symphony, one cannot distinguish between the musical 'idea' and the actual realization of it in sound, a muted solo double bass playing in an unconventional and awkward high register and projected over a drum ostinato.

The onstage–offstage acoustic already adumbrated in the closing stages of *Das klagende Lied* is further explored in the works of these years, for instance in the delicate spatial and dynamic organization of different instrumental groups in the introduction to the First Symphony, or the combination (in different rhythms) of onstage–offstage orchestras in the Second Symphony's finale, and the cadenza (the 'Last Trump' episode) in the same movement. In Mahler's later works, acoustic space is explored within a single orchestra rather than through the contrast of physically separated groups (though the Eighth Symphony significantly revived the earlier practice and the slow movement of the Sixth Symphony had called for offstage cowbells). Extremely subtle dynamic markings and individuating orchestration are two of the means Mahler employs; these allow contrast and mobility between differently constituted bodies of sound within the large orchestra, as with the chamber orchestra which gradually emerges in the Ninth Symphony's first movement (bars 376ff) while still

129

forming part of the main arena of musical action. While some of this may be attributed to the influence of a like-minded innovator such as Berlioz, and also, especially in the Third Symphony, to Mahler's experience of open-air music (as with Ives), the principal generator of many of his acoustic experiments, most of which at this stage were of an explicitly dramatic character, was the opera house.

The influence of theatrical music extends even to Mahler's use of tonality, as with the progressive key schemes in the symphonies which clearly have their roots and logic in dramatic or narrative ideas. Clear examples of this are the G–E of the Fourth Symphony (earthly life–Paradise or experience–innocence) or the more orthodox C minor–E♭ of the Second Symphony (death–resurrection). But it is even apparent in the symphonies which seem superficially to be concentric in tonality, as in the First Symphony where the final affirmation of D has to be fought for (this is part of the narrative of the work: the Hero wins through). In the Third the ultimate D major marks the evolution of the work to an entirely new dramatic (and thus new tonal) plane; although the first movement opens with a substantial slow introduction in D minor, the body of the movement is in F (the key in which Mahler for a long time designated the work), and the D major of the finale represents not only the conquering of the D minor of ice-bound winter, the dramatic topic of the first movement's introduction, but also the decisive shift from biological evolution (the animation of inert matter in the first movement) to a final manifestation of Divine Love (the Adagio).

In this period Mahler was an unambiguously tonal

composer, though he often used tonality in a very fresh way. There is certainly no feeling of exhaustion about his language, in part because he was working naturally in a style that was often unabashedly diatonic, with occasional forays into highly chromatic harmony. The First Symphony is an admirable example of his uninhibited use of the resources of musical language, which, in the 1880s and later – and this was undoubtedly Mahler's historical good fortune – might still include clearcut diatonic melody and the richness of chromatic (i.e. Romantic) harmony. As it had been for Wagner, so it was still for Mahler a valid juxtaposition; and only later did his tonality diminish in stability, with a pervasive chromaticization of his harmonic style. This trend, though, is already apparent in the Fourth Symphony, which for all the diatonic character of some of its materials, sometimes moves into an elaborately chromatic motivic polyphony where tonality becomes exiguous; compare, for example, the sturdy diatonic tune that is the first movement's second subject (ex.4) with the contorted chromaticism of ex.5, from the development, where a strong sense of tonality has been temporarily suspended. Coherence here derives not from tonal direction but from the horizontal and vertical disposition of motifs drawn from the movement's principal melody and reassembled in continually changing orders and textures, one of Mahler's main constructive processes (cf ex.5, which is exclusively assembled, both

Ex.4 Symphony no.4, 1st movt, 2nd subject

Breit gesungen

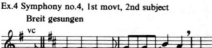

131

Ex.5 Symphony no.4, 1st movt, development

vertically and horizontally, from motifs having their origin in ex.2*a* and ex.4).

Even when Mahler was most selfconsciously a symphonist, as in the first movement of the Second Symphony, he did not feel compelled to match his 'classical' sonata structure with a comparable 'classical' tonal plan. The second subject of the C minor movement moves to the remote key of E major because the drama of the movement required a move not, conventionally, to the relative major but to another tonal world altogether, representing an escape from or alleviation of the preoccupation with death. It was these dramatic considerations that shaped Mahler's tonal strategy, even when outwardly he was at his most formally orthodox. On the other hand, in this same movement he built up such a massive dominant preparation for the return of the tonic at the reprise as to revitalize dynamically this traditional formula. However, when embarking on an orthodox affirmation of the tonic at a reprise, he saw no obligation to recapitulate the expected first subject. In the First Symphony's first movement, for example, D major makes its expected reappearance, but the theme that it affirms is not the first subject (which comes later) but an idea from the beginning of the development. The possibility of this separation of tonal restatement from melodic restatement is a characteristic example of Mahler's fresh way of imagining time-honoured formulae and finding in them new potentialities.

CHAPTER NINE

Middle-period works

In his Fourth Symphony Mahler wrote himself out of his 'Wunderhorn' period but he retained and always drew from the dazzling array of techniques that he had assembled. The next period, from 1901–2 onwards, saw a disciplining of those techniques and the application of them to large-scale instrumental forms, dramatic still in organization, but in a new way and without recourse to song, to song-based movements or to programmes. The narrative principle, as strong as ever, became totally internalized, self-explanatory in purely musical terms and above all in the meaning to be deduced from the broad musical shapes in which Mahler worked. The fact that the Artist as Hero was no longer projected through programmes or through *Wunderhorn* or other texts, involved the development of an even stronger sense of autobiography in the music he wrote after 1901: he became, increasingly, the programme of his own symphonies.

The decisive shift in direction after the long 'Wunderhorn' period, when Mahler had so freely and colourfully shuffled between the modes of symphonic poem and symphony, does not mean that he abandoned the orchestral song or song cycle. On the contrary, just as the *Gesellen* cycle and *Wunderhorn* songs were directly related to the whole creative period that followed, so the last two *Wunderhorn* settings, the

Kindertotenlieder and the Rückert songs, bear a significant relationship to the symphonies surrounding them. At this time, inevitably, there was a decrease in Mahler's dependence on models other than his own music, a natural consequence of his maturity and his stock of resources and experience. The Rückert songs not only epitomize a delicate personal lyricism that was a growing trend in Mahler's art at this time, but may also be regarded technically (especially because of their contrapuntal character, with the voice functioning as another instrumental line) as preliminary studies for *Das Lied von der Erde*. Compare, for instance, ex.6 (from *Um Mitternacht*) with ex.7 (from *Das Lied*). In these songs, in the *Kindertotenlieder* (which Mahler conceived for an authentic chamber orchestra, no doubt with Berlioz's *Nuits d'été* as a precedent) and in the last two *Wunderhorn* settings (*Revelge* and *Der Tamboursg'sell*), which are built on a large scale that reflects the degree to which symphony had infiltrated and fertilized song, there are everywhere relationships between the songs and the parallel symphonies. Examples are the second song of the *Kindertotenlieder* and the Adagietto of the Fifth Symphony; the final song of the same cycle and the introduction to the finale of the Sixth; *Ich bin der Welt abhanden gekommen*, a Rückert song, and the Adagietto of the Fifth; the massive march of *Revelge* and the first movement of the Sixth.

Although the word 'trilogy' is often used in connection with the Fifth, Sixth and Seventh Symphonies, they represent three different approaches. The Fifth is in three parts and five movements, the whole enclosed within a tonal scheme (C♯ minor–D, i.e. leading note–tonic) symbolizing a built-in dramatic narrative quite as

Ex.6 *Um Mitternacht*

unambiguously as the G–E of the Fourth. A surprising
feature of this stage, when Mahler was preoccupied with
integral, large-scale instrumental composition, was the
order in which the movements were composed. It is
understandable that the works from the 'Wunderhorn'
years evolved in the process of composition rather than
fulfilling a predetermined pattern. But the later symphonies, too, though certainly not assembled in so
ad hoc a manner as some of their predecessors, were by
no means composed in the order of the four- or five-
movement sequences that were usually their ultimate
shape. The big central Scherzo of the Fifth was the first
movement of the symphony to materialize, and Mahler
envisaged it as depicting his Hero at the height of his
powers. Though the symphony has no defined programme, a clear narrative runs from the opening funeral

Ex.7 *Das Lied von der Erde*, 'Der Einsame im Herbst'

march through the ensuing agitated Allegro to the pivotal Scherzo, in which the Hero emerges from Part 1, if not precisely unscathed, then at least able in the Scherzo (Part 2) to show the vitality and confidence that makes the triumphant conclusion of Part 3 a logical dénouement. As the musical weight of the symphony rests as

12. *Mahler (extreme left) at a party in the garden of the Moll villa, with Max Reinhardt, Carl Moll and Hans Pfitzner*

much in the mid-point as in the finale, the formal problem that Mahler faced was avoiding anticlimax in the
finale if he allowed the Scherzo to resolve too many
conflicts too soon. This meant that the Scherzo had to
function as a genuine pivot, a hinge on which the drama
could turn from the dark to the light. On this occasion
Mahler was more prudent in his scaling of the opening
funeral march than he had been in the Second
Symphony. There the sheer weight and mass of the
musical invention and the grandeur of the form had
made it excessively difficult for him to continue the
work until he was rescued by the programmatic idea of
resurrection. The much less complex form of the Fifth's
first movement, a march and two trios, while not lacking
in weight, makes a suitably dramatic opening for a Part
1 that is all 'Sturm und Drang', but it skilfully leaves
open developmental possibilities taken up by the ensuing
Allegro which in turn leads on to the Scherzo.

It was essential that the Scherzo should not be too
carefree if the finale were not to be deprived of its
proper role; and Mahler maintained the Scherzo's
dramatic tension by an ingenious mixture of two dances,
the ländler and the waltz. The rustic ländler is exuberant
and optimistic, while the sophisticated waltz, though
starting out innocently, is unexpectedly subjected to the
impact of worldly experience, and more particularly to
the disintegrating, distorting effect of Mahler's ironic
mode (see nos.15–17 in the score).

The fourth movement, the Adagietto for harp and
strings, is not an isolated stretch of slow music but a
slow introduction to the rondo finale, into which it leads
without a break. The materials of the Adagietto – the
contours of whose melody (see ex.8*a*) are representative

Ex.8
(a) Symphony no. 5, Adagietto

(b) Symphony no. 6, 1st movt, 2nd subject

(c) Symphony no. 7, 1st movt, 2nd subject

of the new lyric vein which breaks surface in Mahler's music at this time – are incorporated as episodes into the finale and subjected to all manner of variations. It is the interpolating of the Adagietto into the exuberantly contrapuntal finale that provides the inner tension and contrast which are at last resolved by the clinching statement of the chorale (see nos.32–4), which has made two symbolic and unsuccessful attempts to assert itself

(d) *Das Lied von der Erde*, last movt, 'Der Abschied'

(e) Symphony no. 9, opening of Adagio

as early as the second movement (see nos.17–18 and 27–30).

The Fifth illustrates the kind of fresh and convincing formal solutions Mahler found when he turned away from the free exercise of his fantasy in the 'Wunderhorn' symphonies, where the word could always be used for clarification, and confronted instead the disciplines and challenges of the purely instrumental symphony. The Fifth also shows how important the ideas of a programme and dramatic continuity remained for him, even though now implicit. It also initiated a new period in which, paradoxically, an acknowledged orchestral

141

master laboured unceasingly, often for years, to achieve precisely his ideal instrumental sound for each work. From the Fifth Symphony there is a shift away from the wind-based sonority of the 'Wunderhorn' symphonies towards an orchestral sound embodying a more vivid string presence, a development from which arose a wholly characteristic string music typified by the Adagietto. Ex.8 includes examples of the diverse kinds of emphatically string-based, often appoggiatura-laden melody that Mahler produced in his middle and final periods.

The narrative principle retains its significance for the understanding of the ensuing symphonies. If the Seventh has remained the problem work of the 'trilogy' – it has won nothing like the esteem in which the Fifth and Sixth are generally held – this may be attributed to the relative inaccessibility and indistinctness of its programme, of which a clear perception is necessary if this fascinating but somewhat wayward work is to be successfully experienced. There has always been an association of the symphony with nocturnal nature, not least through the title 'Nachtmusik', which Mahler gave his two slow movements. Indeed, the composer's programmatic hint has led to the use of an illicit descriptive title 'Lied der Nacht' for the work as a whole; the inadequacy of this is brought home by the finale, one of the brightest and least nocturnal of movements in any of the symphonies. It is this movement that has sometimes seemed a stumbling-block to listeners, because it is such an uncompromising contrast to what has preceded it. But here perhaps there is an aid to comprehension in another authentic clue dropped by the composer, who seems to have thought of the finale as the equivalent of sunlight after darkness and to have referred to it on occasion as 'Der Tag'. This, coupled with fresh research and the recovery of further recollections of what Mahler himself had to say about his Seventh, allows the poetic strategy governing the unfolding of the five movements to be defined with some certainty; the first four express various intensities of light and shade in nocturnal perspective until the last vestige of night is dispelled by the finale, but not before Mahler has reintroduced material from the first movement to remind his audience that without darkness there can be no triumph of light (the transformed affirmation of the first movement's Allegro main theme in the last bars is especially convincing in this respect).

The Seventh, then, discloses a continuity close to the narrative principle encountered in its predecessors, a poetic scheme reinforced, as so often before, by a progressive tonal organization, in this case from the B minor/E minor–major of the first movement to the C major of the finale. The first movement also shows a continuity of a different kind. It is a huge sonata structure which in many ways is a counterpart to the first movement of the Sixth. However, it explores a sound-world quite its own, combining Straussian luxuriance and Mahlerian austerity in a bony texture which exploits chains of 4ths, both vertically and horizontally (ex.9, p.144). In the context of what must now be considered the work's established links with nature, it is interesting to note that the rhythm for the slow introduction to the first movement came to Mahler when – or rather, through – rowing on a lake, in itself a timely reminder, like the movement itself, that the composer's relation to nature was not stationary and contemplative

Ex.9 Symphony no.7, 1st movt

but rhythmic and energetic (march rhythms are as conspicuous in the Seventh as they are elsewhere).

The first movement, released by Mahler's oars, was
in fact the last to be completed. For the middle part of
the symphony he conceived a triptych of contrasted
nocturnes, the two slow movements, each entitled
'Nachtmusik', framing a shadowy, spectral Scherzo, in
which the disintegration of the image of the waltz is

carried even further than in that of the Fifth. Thus the symphony, from its brooding B minor introduction, progresses through various shades of darkness. The blackest and bleakest moment is reached at mid-point. The slow march (the first of the two serenades) is 'fantastic' in the old 'Wunderhorn' manner, though carried through with all the orchestral subtlety and sophistication of Mahler's maturity. In fact, it is a late *Wunderhorn* song, but for orchestra alone, a characteristic example of how Mahler was now able to choose models from his own music and regenerate them in often surprising ways. The second 'Nachtmusik' (Andante amoroso) is Mahler's equivalent of a Schumann characterpiece, a kind of extended Albumblatt for a large and idiosyncratic orchestra (mandolin and guitar are prominent).

As for the rondo finale, its shock-tactics can now be seen to be part of the programmatic intent and heard, in all its C major radiance, as a logical outcome of the symphony's opening B minor, in the same way that C♯ minor in the Fifth inexorably results in a concluding D major. Debate about the finale is likely to continue, but there can be no denying its individual features, for example its incessant rhythmic variation, an emphatic diatonicism (its 'white-note' character is doubtless bound up with its poetic 'daylight' function) and its bold, unmodulated tonal switches (cf the double bar before no.230).

The Seventh Symphony sustains the innovatory pattern of the Fifth while at the same time reflecting, in its first movement, the formal preoccupations of the Sixth. The Sixth Symphony is indeed Mahler's most 'formal' symphony, despite the fact that it is also one of

145

his most personal, emotionally charged creations; it too relies on a clear dramatic organization to underpin its total structure. The Sixth is also Mahler's only 'tragic' symphony, that is, in which the Hero is defeated (elsewhere there is only triumph or resignation as a dénouement). Laid out in the conventional four movements, it is one of the few Mahler symphonies unambiguously to assert a concentric tonality: the work begins and ends in A minor, and the Scherzo is in the same key. The concentration on A minor is as symbolic of the work's drama as any of the progressive tonal schemes elsewhere: the very fact that there is no escape for the implied Hero from the fateful key that represents destiny and death constitutes the drama played out through the work. In the finale, one of the longest and most massive of all Mahler's symphonic movements, the death of the Hero is symbolically enacted in the last of the movement's shattering climaxes.

Many distinguished musicians, including Berg and Webern, have judged the Sixth Symphony to be among the greatest of Mahler's works, possibly because of its remarkable equilibrium between form and drama. Certainly, compared with some of the other symphonies, there is less of the tension between classical and narrative formal concepts. This is because the drama in the Sixth, despite the fleeting turn to A major at the end of the first movement or the repose offered by the slow movement's E♭, continually returns to the tragic situation and the tonic minor established at the outset of the first movement. As a result, the emphatic recapitulatory procedures in the symphony, which are part of its classicism, work hand-in-glove with the leading dramatic idea, the Hero's tragic destiny. It is no less important to recognize that Mahler's radical departures from his classical models, for example the tonal strategy which determines the shape of the seemingly 'orthodox' recapitulation and coda of the first movement, were generated by exclusively dramatic considerations. The drama ultimately has the upper hand even in this, the most 'classical' of his symphonies.

The momentum of the opening march spills over into the finale. Once again Mahler uses precedents from his own music as models, this time bringing to its symphonic apotheosis the military imagery that had been a feature of his work since his earliest days. But it is not only the exaltation of the march that is characteristic of all three middle-period symphonies. There is also the striking development in Mahler's contrapuntal thinking. He had always favoured polyphonic textures; in the First Symphony he had established a kind of motivic counterpoint, especially as a means of development (cf also ex.5 above from the Fourth Symphony). The motivic work is no less intricate in the post-'Wunderhorn' period symphonies: the slow introduction to the finale of the Sixth is an example of Mahler unfolding on an imposing scale the many motifs out of which the ensuing movement and its principal themes are to be built. Yet in the textures of these instrumental symphonies emphasis is on the combination of extended melodies rather than on the virtuoso juggling with motifs. This new aspect of Mahler's technique appears with special force in the finale of the Sixth, which sustains page after page of counterpoint, almost Palestrinian in its breadth (ex.10, p.148).

The Sixth, because of its 'classical' shape, is the only symphony by Mahler in which a slow movement plays its traditional role within a four-movement scheme. Elsewhere there is either no real slow movement at all,

147

Ex.10 Symphony no.6, finale

as in the Seventh Symphony, or the slow movement functions in some other role as well. This may be as a finale (the Third and Ninth), as the introduction to a finale (the Fifth), as a first movement (the Tenth), as a vocal movement (*Urlicht*, in the Second) or as a parody (the First). Even the slow movement of the Fourth plays an explicitly dramatic role in relation to the *Wunderhorn* song finale; for instance the great dramatic outburst of E major is the movement's climax and makes full sense only in the light of the finale's switch to the 'heavenly' key.

CHAPTER TEN

Symphony no.8 and 'Lied von der Erde'

After these instrumental symphonies Mahler wrote two works, the Eighth Symphony and *Das Lied von der Erde*, each of which may be regarded as synoptic in character; that is, each comprehensively sums up and develops on a very large scale, with large resources, musical ideas, forms and media that had been Mahler's long-standing preoccupations. It cannot be assumed that this was a conscious act on his part, but it cannot wholly be chance that it was after the three middle-period symphonies – each one representing a different approach to writing a convincing work without recourse to words – that he returned again to vocal resources, and in part to vocal forms.

The massive, often very complex-textured Eighth Symphony and the often sparse-textured *Das Lied von der Erde* share a common influence of Bach, of first importance in Mahler's later music, and an influence that had already been partly responsible for the increasing emphasis given to contrapuntal textures in his music from the turn of the century onwards. Mahler's familiarity with and admiration of Bach was already evident in the 1890s in the textures of the Scherzo of the Second Symphony (no.38ff). He was profoundly stimulated by his possession of the Bach-Gesellschaft edition, and it is not too much to claim that Bach became for him during his last years an exemplar. From this preoccupation

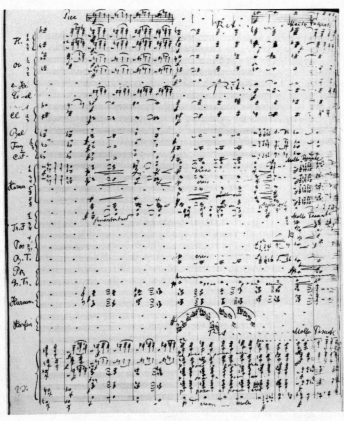

13. Autograph score of a passage from Part 2 of Mahler's Eighth Symphony, composed 1906

also, in 1909, came the assembly of a suite of movements from Bach's orchestral suites, which Mahler selected and edited, and for which he wrote a continuo part.

Symphony no.8, 'Lied von der Erde'

By understanding Bach as a potent, active influence of Mahler's final years much in his late music becomes clear. For example, the key to understanding the first movement of the Eighth Symphony is not by references to 'choral symphony', whether through precedents in Beethoven, Liszt or Mahler himself (the Second Symphony), but by reading the movement as Mahler's tribute to one of Bach's great motets, probably *Singet dem Herrn*, whose vocal polyphony overwhelmed him and which he surely attempted to emulate in his symphony. Rather than a sonata movement (though the sonata model lends another dimension), *Veni creator spiritus* is sensibly approached as a gigantic motet for solo voices, chorus and orchestra – Bach seen through the occasionally distorting lenses of Mahler's creative spectacles. The Eighth Symphony's first movement is a vigorous celebration of the Baroque as Mahler re-imagined it in 1906, an unexpected, fascinating and historic musical event. The sheer mass and ambition of it may be interpreted as a typical late 19th-century conception, but the actual execution is wholly characteristic of Mahler's late choral and orchestral manner.

The second part of the symphony, by way of contrast, is a vast synthesis of many of the forms and media that Mahler had pursued since he first found his voice as a composer. Thus, the setting of the last scene of Goethe's *Faust* represents an amalgam of dramatic cantata, sacred oratorio, song cycle, choral symphony in the manner of Liszt and instrumental symphony, the whole culminating in a final chorale (*Chorus mysticus*) modelled on the concluding chorale of the Second Symphony, though surpassing its precedent in size and ambition.

The Eighth occasionally unleashes torrents of sound, and it was part of Mahler's aesthetic intent to embody his hymn to the redemptive power of love in sonorities of appropriate dimensions. But the vast resources are, characteristically, more often deployed in the most delicate of instrumental effects and subtlest of nuances. Two of many passages in the second part which illustrate the refinement that is as characteristic of the sound of the Eighth as mass or volume are the opening Poco adagio for orchestra and the extraordinary transition scored for an idiosyncratic chamber group of instruments (piccolo, flute, harmonium, celesta, piano, harps and isolated harmonics from a solo string quartet) which leads into the hushed beginning of the *Chorus mysticus*. It is an unusual texture at the lowest dynamic level in what has come to be known misleadingly as the 'Symphony of a Thousand'.

In the scale of its aspiration and its position in Mahler's output, it may fairly be claimed that the Eighth Symphony represents a crowning achievement. The same may be said less controversially of the next work, the song cycle *Das Lied von der Erde*. Mahler described the new work as a symphony, but the ambiguities and uncertainties that surrounded his use of this title with hybrids like the First or Third Symphonies are not relevant here: *Das Lied von der Erde* sums up and fulfils all the symphonic potentialities in the song and the song cycle that he had always seen and explored. Some of the songs in *Das Lied von der Erde* clearly dilate the developmental principles already foreshadowed in the *Gesellen* songs and *Kindertotenlieder*, while at the same time using as models the contrapuntal textures that are a feature of the independent Rückert songs. Others, 'Der Abschied' above all, demonstrate through a remarkable process of integration and juxtaposition the kind of fruitful symbiosis that Mahler achieved in this final stage of his creative life. The central development section of 'Der Abschied', for example (nos.36–48), is not only a big 'song' for orchestra alone, but also a rigorous development of some of the principal motifs out of which the whole closing song is built. There is nothing more symphonic in Mahler's instrumental symphonies.

As a summing up *Das Lied von der Erde* shares some common ground with the Eighth Symphony; in many other respects the song cycle is the symphony's direct opposite. Mass and volume of sound give way to a preoccupation with the most refined and sparest of orchestral textures. It was a natural enough change in emphasis, since the relationship of solo voice to orchestra is different from that of massive choral forces to orchestra; and while the symphony is the most public of Mahler's works, *Das Lied* is one of his most personal. From the Eighth Symphony onwards autobiography replaced the narrative principle that had previously been so important an organizing principle for Mahler, even in his middle-period symphonic projections of the Artist as Hero. The metamorphosis that had begun effectively with the composition of the Fifth Symphony (1901–2) was to be completed in the series of final works initiated by the Eighth in 1906. Here, he attempted something novel, a bold effort to unify a work through the expression of a personal philosophical idea, that is, to relate the idea of human love, Eros, and its redemptive power (the topic of the second part of the symphony) to both the creative spirit which fires the artist and the Creator,

153

God, who endows the artist with creativity (the philosophical substance of the first part).

This departure from the narrative principle is also apparent in the concentric tonal scheme of the Eighth Symphony, which asserts an unshakable and enclosing E♭, with none of the traumatic implications of the Sixth's all-embracing A minor. It is significant that in Mahler's final phase concentric or non-dramatic key schemes prevail. The Tenth Symphony is focussed on F♯ major; *Das Lied von der Erde*'s A minor–C frame is free of the evolutionary drama implied by the identical relationship in the Second Symphony (C minor–E♭); and the Ninth proceeds from a D unstable in mode to the D♭ of the finale. The old dynamic concept of a coordinated, developing tonal scheme tied to a clear narrative line has altogether disappeared. The new emphasis is on the primacy of personal experience; and each large tonal scheme must be independently assessed and interpreted.

But although the former narrative principle was less pronounced in Mahler's late phase, there was instead a rather differently formulated cross-movement continuity, bound up with the resolution, at the end of a work, of a conflict, tension or duality stated at the outset. Resolution had had a role as part of the narrative process but now it became virtually a principle in its own right. In *Das Lied*, for example, the entire work is dedicated to reconciling the conflict embodied in the principal poetic or quasi-dramatic idea which is the topic of the first song: the confrontation of life and death. In the Ninth the finale resolves – or pacifies – the explosive tension that generates the first movement; in the Tenth the finale reintroduces and then, at its close, mitigates and transforms the famous and massive vertical dissonance which so disconcertingly ruptures the texture of the opening Andante–Adagio, a passage that crosses the boundary between intense expressiveness and expressionism (cf bars 203–8 of the first movement with bars 275–83 of the finale).

When this basic antithesis is absent, as it is largely (though not entirely) from the Eighth Symphony, and there is no conflict to resolve, one might think the consequences would not be to the composer's advantage, especially from a formal point of view, since the most interesting and innovatory of Mahler's forms in his later years represented a conspicuous duality. This is particularly true of Mahler's conception of the strophe, the unit fundamental to his songs, which in works like the *Kindertotenlieder* and *Das Lied von der Erde* he used with unprecedented subtlety and an intricacy (for example in the first of the *Kindertotenlieder*) which can only be described as symphonic in its elaboration. The treatment of the strophe in the two late song cycles clearly brings them into the context of the surrounding symphonies; at the same time there are movements in the symphonies (the first movement of the Ninth, for instance) which reflect Mahler's preoccupation with strophic form. Moreover, the form of the individual strophe often reveals in its detailed construction the duality that is the mainspring of the whole work. The dichotomous strophe that is the basis of the first song of the *Kindertotenlieder* provides the model for the later song cycles and symphonies. Virtually every dimension of it embodies a duality: minor and major modes (D minor and D major), contrasted character of invention, contrasted orchestral textures (woodwind/strings) and

155

contrasted poetic images (darkness/light). There is no clearer example of the method and symmetries of Mahlerian duality.

Although the expression of personal experience may be the unifying feature of this last period, there are common techniques; and, again, the influence of Mahler's study of Bach is often prominent. He was not to repeat the busy, abundant vocal polyphony of the Eighth's first movement, but in both *Das Lied von der Erde* and the Ninth Symphony, Bach's influence recurs in textures and forms that could hardly have been foreseen. The unmeasured, asymmetrical 'prose' style of the statuesque recitatives of 'Der Abschied', with their extraordinary woodwind obbligatos, has behind it Mahler's profound response to the recitatives of Bach's church cantatas and Passions rather than his daily familiarity with operatic recitative. In the Ninth Symphony, too, there are passages like the austere contrapuntal pages in the Adagio finale (e.g. bars 28–48), which embody these new contrapuntal preoccupations; and there is even an entire movement, the Rondo–Burleske, in which the motivic counterpoint that had long been a feature of Mahler's compositional method is consummated.

CHAPTER ELEVEN

Symphonies nos.9 and 10

Not all the technical innovations of *Das Lied von der Erde* were followed up in the later symphonies, but the final works show Mahler again trying to create a convincing shape for the instrumental symphony, the individual form of which, in this last phase, was tailored to the personal narrative that his art had become. The Ninth and Tenth Symphonies are a conscious return to the old formal concerns that had preoccupied him in the 'trilogy' of middle-period symphonies, but with a fresh mobilization of all his imaginative and intellectual powers. The Ninth Symphony was the fourth of Mahler's experiments in four-movement shapes, but has virtually no formal tactic in common with its predecessors, the First (in its final version), Fourth and Sixth Symphonies. If it owes a formal debt at all, then it is to *Das Lied von der Erde*; the Adagio of the Ninth is clearly another kind of 'Abschied', though this time conceived in terms of instrumental symphony.

From almost every other point of view, however, and especially in the first movement, the Ninth Symphony breaks new ground. It even makes some fresh approaches to filling out the frame provided by the first movement and finale. The Scherzo and Rondo–Burleske, considered from one angle, are genuine extensions of the character movements that are a feature of Mahler's symphonic landscapes, but to each example in

157

14. Gustav Mahler with his daughter Anna, Toblach, 1909

the Ninth he brought fresh and formidable thinking. The Scherzo is the most comprehensively ironical of all his ländler-based movements. The ländler is not used as an instrument of irony, but as a subject for irony, and is indeed, finally, ironically dispatched (see no.27ff). The Rondo–Burleske introduces a new type of character movement altogether, a big stretch of fiercely elaborate counterpoint, elaborately orchestrated, as sophisticated

and complex as the preceding Scherzo is at times unequivocally primitive. What unites these two movements
is their common satirical intention. For the Rondo is as
satirical in its sophistication as the Scherzo is in its
simplicities, and these central movements present two
faces of satire, delineated in materials and manipulated
in techniques that have very little in common. There is
no clearer instance than this of Mahler trying to make
unity out of paradox.

It was Berg who referred to the first movement of the
Ninth Symphony as 'the most glorious [Mahler] ever
wrote' and who, in the same letter (undated, but
probably from autumn 1912) offered a clue to the
analysis of this complex Andante comodo. Berg did not
touch on such matters (central though they are to the
conception of the movement) as the juxtaposition of D
major and minor, or the alternation of contrasted
tempos, the Andante of the first (major) section (bars 6–
26) always quickening into the quasi-Allegro of the
(minor) second section (bars 27–46). Nor did he comment on the fact that in his Ninth and Tenth
Symphonies Mahler developed a new type of slow first
movement distinct from a slow movement placed as first
movement. Thus, the Andante of the Ninth, for all its
basically slow pulse, develops the pace, the dramatic
quality and complexity of organization associated with a
fully-fledged symphonic first movement. But Mahler
achieved none of this through orthodox formal gestures,
and though thousands of analytic words have been ex-
pended in the effort to relate the first movement of the
Ninth to the traditional concepts of sonata or sonata-
rondo, it was Berg in 1912 who stumbled on the key to
the real form of the movement when he wrote, 'The

15. *Autograph draft orchestral score from the third movement
of Mahler's Ninth Symphony, composed 1908–9*

whole movement is based on a premonition of death which constantly recurs ... that is why the tenderest passages are followed by tremendous climaxes like new eruptions of a volcano'. It is not Berg's perception of the inner drama that need be stressed but his perception of the dynamic principle around which the whole vast movement is built, that is, the bold idea of a recurring, expanding and exploding crescendo (bars 27–46, 80–107, 174–203, 211–42, 284–316, 372–402) – Berg's volcanic eruptions. The penultimate, longest and loudest of the crescendos gives rise to the extreme disintegration of the movement's materials, and the final reassembly provides a reconciliation of all the disparate components that have been involved. It is neither sonata nor rondo which is the key to the organization of the Ninth's Andante, but the audacious idea of organizing a large and complicated instrumental movement around a dynamic principle, specifically the crescendo.

The pattern of the Tenth Symphony shows Mahler launching out on yet another vast symphonic enterprise, again a five-movement shape (cf the Fifth and Seventh Symphonies) organized as (Andante) Adagio; Scherzo I (Allegro); 'Purgatorio' (Allegretto moderato); Scherzo II (Allegro); Finale (slow introduction–Allegro moderato). Deryck Cooke's reconstruction of a performing edition from Mahler's sketch makes available his last inspirations in their complete context, even though his final version would inevitably have been very different. It also clearly indicates (as Cooke suggested) that Mahler had moved on, spiritually, from the resignation of *Das Lied von der Erde* and the Ninth Symphony to a reconciliation in the Tenth, which a decision during work on the symphony makes explicit: he had at one stage considered bringing the work to a close in B♭ but finally settled for a coda to the last movement that unequivocally affirms the tonic, F♯, in the major mode. Yet even while the significance of this assertion of concentric tonal organization in the Tenth is acknowledged, it is also important to remember that Mahler's dramatic key schemes were not confined to the large-scale organization of a work but were also a prominent aspect of the tonal organization within movements. The outwardly concentric Tenth itself reveals interesting instances of this inner dramatic organization, such as the fourth movement's regress from E minor to D minor and the fifth's progress from D minor to F♯.

In the works from Mahler's final phase principles from the past continue to form the basis of his methods of organization, but it is the freshness of inflection that is of particular importance. The Tenth Symphony may share its concentric tonality with the Sixth or the Eighth, but now the achievement of a serene F♯ major is evidence of the resilience of an individual spirit, not the outcome of an externalized, objectified drama. Again, the regressive key scheme, D–D♭, of the Ninth Symphony may seem another example of Mahler's tonal symbolism, and so from one point of view it is. But here the falling back by a semitone was not conditioned by the pre-existence of a narrative scaffold, as in the Fifth, which presents the reverse tonal shift (C♯ minor to D), but by Mahler's response to his own destiny.

Although the Tenth Symphony was not completed, it is possible to discern many new stylistic and formal features. There is certainly no slackening in intensity; and it was perhaps the unleashed force of the expressive intent that gave rise to the huge melodic leaps and skips of the opening Adagio. But Mahler's constructive powers are no less evident than his intensity of feeling. It is interesting, indeed, how rationally (from a technical point of view) he treated such highly emotive material as the great Adagio theme itself, for instance the inversion of the theme at bar 69 (ex.11), a practice comparable with one aspect of the strict manipulation of a note row. Given the intensely chromatic nature of the theme, it is not surprising that he should have adopted this technique as a means of variation and development which at the same time preserves the basic shape of the melody.

Ex.11 Symphony no.10, Adagio

The clearest evidence, however, of Mahler's undiminished creativity in his last symphonies lies in his continued exploration of large forms; and once again it is to his first movements and finales that he brought all his powers of renewal and his pioneering spirit. The first movement of the Tenth shows him consolidating the new ground that had already been opened up in the first movement of the Ninth. It seems clear that the finale of the Tenth would have been counted among the most remarkable of Mahler's finales. It is the most extensive and comprehensive in recapitulatory gestures, which represent not so much restatement as development and the discovery of fresh potentialities in material from all

163

the preceding movements. This new conception of the cyclic finale singles out the last movement of the Tenth as a special achievement which fittingly rounds off a lifetime of formal innovation.

WORKS

Edition: *G. Mahler: Sämtliche Werke: kritische Gesamtausgabe*, ed. Internationale Gustav Mahler Gesellschaft (Vienna, 1960–) [MW]

This list includes only extant works and lost works whose music survives in other compositions; only the principal manuscripts of complete works are given, except where noted. For keys, an oblique stroke denotes alternation, a dash denotes a progressive scheme.

Numbers in the right-hand column denote references in the text.

* – autograph MS

Title, Key, Forces	Date	First performance, Publication	Remarks, MS, Edition	
Piano Quartet, a, inc.	?1876–8	?1876–8: New York, 12 Feb 1964 [ed. D. Newlin]; ed. P. Ruzicka (Hamburg, 1973)	movt 1 and frag. [24 bars] of Scherzo only; *US-NYpm	82
Rübezahl (Mahler), opera	?1879–83	—	music lost; lib, NH	84–5, 86, 89
Lieder (Mahler), T, pf		Radio Brno, 30 Sept 1934	from projected set of 5 songs; *Mahler-Rosé Collection, U. of Western Ontario	85
1 Im Lenz, F: Ab	19 Feb 1880			
2 Winterlied, A F	27 Feb 1880			
3 Maitanz im Grünen, D	5 March 1880			
Das klagende Lied (Mahler), cantata [c–a], S, A, T, mixed vv, orch	1878–80, rev. 1892–3, 1898–9	pubd version, Vienna, 17 Feb 1901; orig. version [a–a], incl. Waldmärchen, Radio Brno, 28 Nov 1934 [in Cz.], Waldmärchen only, 2 Dec 1934 [in Ger.]; orig. version, complete, Vienna Radio, 8 April 1935 [in Ger.] (Leipzig and Vienna, 1902) [orig. pts.2 and 3]; Waldmärchen (New York, 1973)	early version of Hans und Grethe orig. in 3 pts. Waldmärchen, Der Spielmann, Hochzeitsstück; orig. version copy, NH; *1st rev. version, NYpm; MW xii	84, 85, 86, 89, 113–17, 118, 124, 127, 129
Lieder und Gesänge, v, pf		(Mainz, 1892)	later renamed by publisher Lieder und Gesänge aus der Jugendzeit; *Mahler-Rosé Collection, U. of Western Ontario	85, 94, 97, 115
i	1880–87			
1 Frühlingsmorgen (R. Leander), F		Budapest, 13 Nov 1889		115
2 Erinnerung (Leander), g–a		Budapest, 13 Nov 1889		86
3 Hans und Grethe (Mahler), D		Prague, 18 April 1886	reworking of Maitanz im Grünen	86, 115
4 Serenade aus Don Juan (Tirso de Molina, trans. L. Braunfels), Db		? Leipzig, Oct 1887		85, 115
5 Phantasie aus Don Juan (Tirso de Molina, trans. Braunfels), F#/b		? Leipzig, Oct 1887		

Title, Key, Forces	Date	First performance, Publication	Remarks, MS, Edition	
ii (from C. Brentano and A. von Arnim: Des Knaben Wunderhorn)	1887–90			93–4, 97
1 Um schlimme Kinder artig zu machen, E		Munich, 1899–1900 season	orig. poem title Um die Kinder still und artig zu machen	
2 Ich ging mit Lust durch einen grünen Wald, D		Stuttgart, 13 Dec 1907 [?also earlier]	orig. poem title Waldvögelein	
3 Aus! Aus!, Db		Hamburg, 29 April 1892	orig. poem title Abschied für immer	
4 Starke Einbildungskraft, Bb	1887–90	Stuttgart, 13 Nov 1907[? also earlier]		94
iii (from Des Knaben Wunderhorn)				
1 Zu Strassburg auf der Schanz, F#/#-B/b		Helsinki, Nov 1906 [? also earlier]	orig. poem title Der Schweizer	
2 Ablösung im Sommer, db		Berlin, 1904–5 season	orig. poem title Ablösung	120
3 Scheiden und Meiden, F		Budapest, 13 Nov 1889	orig. poem title Drei Reiter am Tor	
4 Nicht wiedersehen!, c		Hamburg, 29 April 1892		
5 Selbstgefühl, F		Vienna, 15 Feb 1900		
Lieder eines fahrenden Gesellen (Mahler), song cycle, low v, orch/pf	Dec 1884–1885, rev. ?1891–6	Berlin, 16 March 1896 [with orch]; ? perf. earlier with pf; orch, pf versions (Vienna, 1897)	orchd ?1890s; several discrepancies among versions; • pf version Mahler-Rosé Collection, U. of Western Ontario; •early full score [? 1891–3], Mengelberg-Stichting, NL-DHgm; vocal score ed. C. Matthews and D. Mitchell, 1977; vocal score, MW xiv	89, 114, 117–18, 125, 134, 152
			text based on poem from Des Knaben Wunderhorn, Wann mein Schatz	
1 Wenn mein Schatz Hochzeit macht, d–g				
2 Ging heut' morgens übers Feld, D F#				
3 Ich hab' ein glühend Messer, d–eb				
4 Die zwei blauen Augen, e–f				
Der Trompeter von Säkkingen (J. V. von Scheffel), incidental music, orch	1884	Kassel, 23 June 1884	lost; ? 1st no. Ein Ständchen am Rhein, used as Andante (Blumine), Sym. no.1, orig. version	118
Symphony no.1, D, orch	?1884–March 1888, rev. 1893–6, rev. 2/ c1906	Budapest, 20 Nov 1889; (Vienna, 1899, rev. 2/1906) [4 movts]; Blumine (Bryn Mawr, Penn., 1967)	orig. called Sym. Poem, later 'Titan'; orig. in 5 movts, Andante (Blumine) discarded in final rev.; •[5 movts] US-NH; MW i [4 movts]	118, 89, 93, 96, 97, 98, 101, 113, 115, 118–19, 122, 124, 129, 130, 131, 133, 147, 152, 157

Work	Date	First performance	Remarks	Refs
Symphony no.2, c–E♭, S, A, mixed vv, orch	1888–94, rev. 1903	movts 1–3, Berlin, 4 March 1895; complete, Berlin, 13 Dec 1895; (Leipzig, 1897, rev. 2/1903); arr. 2 pf (Leipzig, 1895)	movt 4 text (Urlicht) from Des Knaben Wunderhorn; movt 5 text Klopstock, Mahler; *NYpm*; copy with changes, *NH* [less movt 4]; MW ii	93, 94, 98, 101, 107, 115, 118, 121, 122, 127, 128, 129, 130, 133, 139, 141, 142, 148, 149, 151, 154
Des Knaben Wunderhorn (Brentano and Arnim), songs, v, pf/orch		pf, orch versions (Vienna, 1899)	orig. called Humoresken; all first pubd separately, except no.11 of which no version for solo v and orch by Mahler exists	98, 120ff
1 Der Schildwache Nachtlied, B♭	28 Jan 1892	Berlin, 12 Dec 1892 [with orch]	orchd by 26 April 1892; *D-B [pf], A-Wgm [orch]	
2 Verlor'ne Müh, A	1 Feb 1892	Berlin, 12 Dec 1892 [with orch]	orchd by 26 April 1892; *D-B [pf], A-Wgm [orch]	
3 Trost im Unglück, A	22 Feb 1892	Hamburg, 27 Oct 1893 [with orch]	orig. poem title Geh du nur hin; orchd by 26 April 1892; *D-B [pf], A-Wgm [orch]	
4 Wer hat dies Liedlein erdacht?, F	6 Feb 1892	Hamburg, 27 Oct 1893 [with orch]	orchd by 26 April 1892; *D-B [pf], A-Wgm [orch]	
5 Das irdische Leben, b♭ [Phrygian]	between April 1892 and summer 1893	Vienna, 14 Jan 1900 [with orch]	orig. poem title Verspätung; *private collection, USA [pf draft], US-NYpm [orch]	
6 Des Antonius von Padua Fischpredigt, c	8 July 1893	Vienna, 29 Jan 1905 [with orch]	orchd 1 Aug 1893; *NYpm [pf], C-A [orch]	
7 Rheinlegendchen, A	9 Aug 1893	Hamburg, 27 Oct 1893 [with orch]	orig. poem title Rheinischer Bundesring; orchd 10 Aug 1893; *D-B [pf], US-NYpm [orch] *NYpm [pf]	
8 Lied des Verfolgten im Turm, d	July 1898	Vienna, 29 Jan 1905 [with orch]	orig. poem title Unbeschreibliche Freude: *NYpm [pf]	
9 Wo die schönen Trompeten blasen, d	July 1898	Vienna, 14 Jan 1900 [with orch]	orig. poem title Wettstreit des Kuckucks mit der Nachtigall; *private collection, USA [pf draft]	
10 Lob des hohen Verstandes, D	between 21 and 28 June 1896	Vienna, 18 Jan 1906 [with pf]	orig. poem title Armer Kinder Bettlerlied; composed for Sym. no.3; pf version, before 1899; *orch draft, NYpm	
11 Es sungen drei Engel, F	11 Aug 1895 [orch draft of sym. movt]	Krefeld, 9 June 1902 [in Sym. no.3]		121
12 Urlicht, D♭	?1892	Berlin, 13 Dec 1895 [in Sym. no.2]	orchd 19 July 1893; used in Sym. no.2	121

Title, Key, Forces	Date	First performance, Publication	Remarks, MS, Edition	
Das himmlische Leben, G–E	10 Feb 1892	Hamburg, 27 Oct 1893 [with orch]	orig. poem title Der Himmel hängt voll Geigen; orchd 12 March 1892; used in Sym. no.4; *D-B [pf]. *A-Wgm [orch]	121, 123
Symphony no.3, d/F–D, A solo, women's vv, boys' vv, orch	1893-6, rev. 1906	movt 2, Berlin, 9 Nov 1896; movts 2, 3, 6, Berlin, 9 March 1897; complete, Krefeld, 9 June 1902; (Vienna, 1899, rev. 2/1906)	movt 4 text from Nietzsche: Also sprach Zarathustra; movt 5 text (Es sungen drei Engel) from Des Knaben Wunderhorn; *US-NYpm; MW iii	98, 99, 101, 115, 120, 121, 122-3,127,130,141,142, 148, 152, 224
Symphony no.4, (b)/G–E, S, orch	1892, 1899-1900, rev. 1901-10	Munich, 25 Nov 1901; (Vienna, 1902, rev. 2/1906)	movt 4 text (Das himmlische Leben) from Des Knaben Wunderhorn; *A-Wgm; MW iv	101, 111, 118, 121, 123-5, 126, 127, 130, 131-3, 134, 136, 141, 142, 147, 148, 157
Symphony no.5, c♯–D, orch	1901-2, scoring repeatedly rev.	Cologne, 18 Oct 1904; (Leipzig, 1904, rev. 2/1904)	*US-NYpm; MW v	103, 106, 135-7, 139-42, 145, 148, 153, 161, 162
Kindertotenlieder (Rückert), song cycle, 1v, orch	1901-4	Vienna, 29 Jan 1905; full, vocal scores (Leipzig, 1905)	*NYpm [pf 2-5; orch 1-5]; *A-Wgm [pf]; MW xiv/5	103, 106, 135, 152, 155
1 Nun will die Sonn' so hell aufgeh'n, d				
2 Nun seh' ich wohl, warum so dunkle Flammen, c				135
3 Wenn dein Mütterlein, c				
4 Oft denk' ich, sie sind nur ausgegangen, E♭				
5 In diesem Wetter, in diesem Braus, d–D				135
Lieder, 1v orch/pf		nos.1-6, Vienna, 29 Jan 1905 [with orch]; nos.1-6, orch, pf versions (Leipzig, 1905)	all first pubd separately; later renamed by publisher Sieben Leider aus letzter Zeit; nos.3-7: 'Rückert-Lieder'	103, 135, 153, 155
1 Revelge (from Des Knaben Wunderhorn), d or c	July 1899		*NYp [orch]	101, 134, 135
2 Der Tamboursg'sell (from Des Knaben Wunderhorn), d	Aug 1901		*NYpm [orch]	
3 Blicke mir nicht in die Lieder (Rückert), F	14 June 1901		*NYpm [orch]; *A-Wn [pf]	103, 134, 135
4 Ich atmet' einen Linden Duft (Rückert), D	July or Aug 1901		*D-Mbs [pf]	
5 Ich bin der Welt abhanden gekommen (Rückert), F/E♭	16 Aug 1901		*owned by H.-L. de La Grange, Paris [pf version]	

Work	Composed	Performance; publication	Sources / notes	Pages
6 Um Mitternacht (Rückert), b	summer 1901		*Schoenberg Archives, Los Angeles [orch, inc]; *A-Wst [pf draft]; La Grange [pf]	135, 136
7 Liebst du um Schönheit (Rückert), C	Aug 1902	Vienna, 8 Feb 1907 [? also earlier]; (Leipzig, 1907); orch version by M. Puttmann (Leipzig, 1916)	composed with pf acc.; *owned by Anna Mahler, London	102
Symphony no.6, a, orch	1903–4, rev. 1906, scoring repeatedly rev.	Essen, 27 May 1906; (Leipzig, 1906, rev. edn. 1906)	*Wgm; MW vi	106, 107, 129, 135, 140, 142, 143, 145–8, 154, 157, 162
Symphony no.7, (b) e–C, orch	1904–5, scoring repeatedly rev.	Prague, 19 Sept 1908; (Berlin, 1909)	*Concertgebouw Orchestra archives, Amsterdam; MW vii	106, 108, 135, 140, 142–5, 148, 161
Symphony no.8, Eb, 3 S, 2 A, T, Bar, B, boys' vv, mixed vv, orch	summer 1906–7	Munich, 12 Sept 1910; vocal score (Vienna, 1910); full score (Vienna, 1911)	pt.1 text (Veni creator spiritus), hymn attrib. Hrabanus Maurus, Archbishop of Mainz [9th century]; pt.2 text from Goethe: Faust, closing scene; *D-Mbs; MW viii	106, 111, 115, 124, 149ff, 162
Das Lied von der Erde (from H. Bethge: Die chinesische Flöte), sym., a–C, T, A/Bar, orch	1907–9	Munich, 20 Nov 1911; vocal score (Vienna, 1911); full score (Vienna,1912)	*US-NYpm; vocal score, private collection, USA; MW ix	108–9, 135, 137, 141, 149, 152ff, 157, 161, 234
Symphony no.9, D–Db, orch	1908–9	Vienna, 26 June 1912; pf 4 hands (Vienna, 1912); full score (Vienna, 1913)	*A-Wn, *US-NYpm; earlier draft of movts 1–3, facs., ed. E. Ratz (Vienna, 1971); MW x	109, 129, 141, 148, 154, 155, 156, 157–9, 160, 161, 162, 163
Symphony no.10, f#/F#, orch, inc.	1910	movts 1, 3, Vienna, 14 Oct 1924; complete perf. version by D. Cooke, London, 13 Aug 1964; movts 1, 3 (New York, 1951); perf. version by D. Cooke (London and New York, 1976)	most *sketches and drafts owned by Anna Mahler, London; facs., ed. R. Specht (Berlin, Vienna and Leipzig, 1924); facs. with addl sketches, ed. E. Ratz [Munich, 1967]; MW xi a [movt 1 only]	103, 111, 148, 154, 157, 159, 161–4

PERFORMING EDITIONS/ARRANGEMENTS

Suite aus den Orchesterwerken von J. S. Bach, orch, hpd, org (New York, 1910): 1 Ouverture [from Suite no.2, b], 2 Rondeau und Badinerie [from Suite no.2, b], 3 Air [from Suite no.3, D], 4 Gavotte no.1 und 2 [from Suite no.3, D] — 109, 150

A. Bruckner: Symphony no.3, pf 4 hands (Vienna, 1880) [? collab. R. Krzyzanowski] — 84

W. A. Mozart: Die Hochzeit des Figaro (Der tolle Tag), vocal score — 112 (Leipzig, 1907)

C. M. von Weber: Die drei Pintos, full score, vocal score, arr. pf solo (Leipzig, 1888) [reconstruction and augmentation of inc. opera] — 92–3, 94

—: Euryanthe, full score, unpubd; new lib by Mahler (Vienna, 1904)

—: Oberon: König der Elfen, vocal score (Vienna, 1919)

Numerous unpubd edns./arrs., incl. works by Beethoven (Str Qt, f, op. 95, Sym. no.9, d), Schubert (Str Qt, d, D819, Sym. no.9, C), Schumann (4 syms.) and Bruckner (Sym. no.5, Bb)

BIBLIOGRAPHY

CATALOGUES, BIBLIOGRAPHIES AND SOURCE STUDIES

O. Keller: 'Gustav Mahler-Literatur', *Die Musik*, x/18 (1911), 369; suppl. by A. Seidl, *Die Musik*, x/21 (1911), 154

Mahler Feestboek (Amsterdam, 1920) [incl. facs. of MSS]

F. Hadamowsky, ed.: *Gustav Mahler und seine Zeit* (Vienna, 1960) [exhibition catalogue]

K. Thompson: *A Dictionary of Twentieth Century Composers* (London, 1973) [with extensive bibliography]

LETTERS, DOCUMENTS AND ICONOGRAPHY

G. Mahler: Interview, *New York Daily Tribune* (3 April 1910)

A. Roller: *Die Bildnisse von Gustav Mahler* (Leipzig, 1922)

A. Mahler, ed.: *Gustav Mahler: Briefe 1879–1911* (Berlin, Vienna and Leipzig, 1924; Eng. trans., ed. K. Martner, 1979)

H. Holländer: 'Unbekannte Briefe aus Gustav Mahlers Jugend', *Neues Wiener Journal* (16 Sept 1928)

——: 'Unbekannte Jugendbriefe Gustav Mahlers', *Die Musik*, xx (1927–8), 807

W. Schuh and F. Trenner, eds.: 'Hans von Bülow, Richard Strauss: Briefwechsel', *Richard Strauss Jb 1954*, 7–88; Eng. trans. (London, 1955)

H. Holländer: 'Gustav Mahler vollendet eine Oper von Carl Maria von Weber: vier unbekannte Briefe', *NZM*, cxvi (1955), 130

F. Bartoš, ed.: *Mahler: Dopisy* (Prague, 1962) [letters]

F. Grasberger, ed.: *Die Welt um Richard Strauss in Briefen* (Tutzing, 1969)

W. Schreiber: *Gustav Mahler in Selbstzeugnissen und Bilddokumenten* (Hamburg, 1971)

H. Moldenhauer: 'Unbekannte Briefe Gustav Mahlers an Emil Hertzka', *NZM*, xxxv (1974), 544

K. Blaukopf, ed.: *Mahler: sein Leben, sein Werk und seine Welt in zeitgenössischen Bildern und Texten* (Vienna, 1976; Eng. trans., 1976 as *Mahler: a Documentary Study*)

S. Wiesmann, ed.: *Gustav Mahler in Vienna* (New York, 1976)

K. Martner and R. Becqué: 'Zwölf unbekannte Briefe Gustav Mahlers an Ludwig Strecker', *AMw*, xxxiv (1977), 287

J. Theurich: 'Briefe Gustav Mahlers an Ferruccio Busoni', *BMw*, xix (1977), 212

B. and E. Vondenhoff, eds.: *Gustav Mahler Dokumentation: Sammlung Eleonore Vondenhoff: Materialen zu Leben und Werk* (Tutzing, 1978); suppl. (in preparation)

Bibliography

E. Klemm: 'Zur Geschichte der Fünften Sinfonie von Gustav Mahler: Der Briefwechsel zwischen Mahler und dem Verlag C. F. Peters und andere Dokumente', *JbMP 1979*, 9–116

R. Stephan, ed.: *Gustav Mahler: Werk und Interpretation: Autographe, Partituren, Dokumente* (Cologne, 1979)

H. Blaukopf, ed.: *Gustav Mahler, Richard Strauss: Briefwechsel 1888–1911* (Munich and Zurich, 1980)

D. Gutmann: *Auguste Rodin: les bustes de Gustav Mahler* (diss., Institut d'Archéologie et d'Histoire de l'Art, U. of Paris, 1980)

K. Martner, ed.: *Selected Letters of Gustav Mahler* (London, 1980)

E. Reeser, ed.: *Gustav Mahler und Holland: Briefe* (Vienna, 1980)

M. Hansen, ed.: *Gustav Mahler: Briefe* (Leipzig, 1981)

H. Blaukopf, ed.: *Gustav Mahler: Briefe* (Vienna, 1982) [based on A. Mahler's edn., 1924]

——, ed.: *Gustav Mahler: Unbekannte Briefe* (Vienna, 1983)

CONTEMPORARY ACCOUNTS AND MEMOIRS

A. Neumann: *Prager Erinnerungen* (Vienna, 1908)

R. Rolland: *Musiciens d'aujourd'hui* (Paris, 1908, 18/1947; Eng. trans., 1915/R1969)

J. B. Foerster: 'Aus Mahlers Werkstatt: Erinnerungen', *Der Merker*, i (1910), 291

E. Decsey: 'Stunden mit Mahler', *Die Musik*, x (1911), no.18, p.352; no.21, p.143

M. Steinitzer: *Richard Strauss* (Berlin, 1911, 4/1911)

J. Stransky: 'Begegnungen mit Gustav Mahler', *Signale für die musikalische Welt* (1911), July

M. Gutheil-Schoder: 'Gustav Mahler bei der Arbeit', *Der Merker*, iii (1912), 165

A. Bahr-Mildenburg: *Erinnerungen* (Vienna and Berlin, 1921)

N. Bauer-Lechner: *Erinnerungen an Gustav Mahler* (Vienna and Zurich, 1923; Eng. trans., ed. P. Franklin, 1980)

A. Rosé: 'Aus Gustav Mahlers Sturm- und Drangperiode', *Hamburger Fremdenblatt* (5 Oct 1928)

L. Karpath: *Begegnung mit dem Genius* (Vienna, 1934)

B. Walter: 'Persönliche Erinnerungen an Gustav Mahler', *Der Tag* (Vienna, 17 Nov 1935)

A. Mahler: *Gustav Mahler: Erinnerungen und Briefe* (Amsterdam, 1940; Eng. trans., 1946, rev., enlarged 3/1975, ed. D. Mitchell and K. Martner)

J. B. Foerster: *Der Pilger: Erinnerungen eines Musikers* (Prague, 1955)

F. Endler, ed.: *Egon Wellesz: Leben und Werk* (Vienna and Hamburg, 1981)

O. Klemperer: *Meine Erinnerungen an Gustav Mahler und andere autobiographische Skizzen* (Zurich, 1960; Eng. trans., enlarged, 1964 as *Minor Recollections*)

A. Mahler: *And the Bridge is Love: Memories of a Lifetime* (London, 1959; from Ger. orig., *Mein Leben*, 1960)

F. Pfohl: *Gustav Mahler: Eindrücke und Erinnerungen aus den Hamburger Jahren*, ed. K. Martner (Hamburg, 1973)

E. H. Gombrich: *Anna Mahler: her Work* (London, 1974) [with essay by A. Mahler and biographical sketch]

D. Schuschitz: *Die Wiener Musikkritik in der Ära Gustav Mahler 1897–1907: eine historisch-kritische Standortbestimmung* (diss., U. of Vienna, 1978)

P. Heyworth: *Otto Klemperer: his Life and Times*, i: *1885–1933* (Cambridge, 1983)

K. Monson: *Alma Mahler: Muse to Genius* (Boston, 1983)

OBITUARIES

Obituary, *Illustriertes Wiener Extrablatt* (19 May 1911)

S. Langford: Obituary, *Manchester Guardian* (19 May 1911)

Obituary, *The Times* (20 May 1911)

Obituary, *Neues Wiener Tagblatt* (20 May 1911)

H. E. Krehbiel: 'Death of Mr. Mahler', *New York Daily Tribune* (21 May 1911)

SPECIAL PERIODICAL ISSUES

Die Musik, x/18 (1911)

Der Merker, iii/5 (1912)

Musikblätter des Anbruch, ii/7–8 (1920)

Moderne Welt, iii/7 (1921–2)

Musikblätter des Anbruch, xii/3 (1930)

ÖMz, xxxiv/6 (1979)

LIFE AND WORKS; GENERAL BIOGRAPHICAL AND MUSICAL STUDIES

M. Graf: 'Gustav Mahler', *Wagner-Probleme und andere Studien* (Vienna, 1900), 122

L. Schiedermair: *Gustav Mahler: eine biographisch-kritische Würdigung* (Leipzig, 1901)

R. Specht: *Gustav Mahler* (Berlin, 1905)

P. Stefan: *Gustav Mahlers Erbe* (Munich, 1908)

P. Stefan, ed.: *Gustav Mahler: ein Bild seiner Persönlichkeit in Widmungen* (Munich, 1910)

P. Stefan: *Gustav Mahler: eine Studie über Persönlichkeit und Werk* (Munich, 1910, 2/1913, 4/1920; Eng. trans., 1913)

R. Specht: *Gustav Mahler* (Berlin, 1913, 18/1925)

G. Adler: *Gustav Mahler* (Vienna, 1916) [orig. in *Biographisches Jb und deutscher Nekrolog*, xvi (1914), 3–41]

A. Neisser: *Gustav Mahler* (Leipzig, 1918)

Bibliography

H. F. Redlich: *Gustav Mahler: eine Erkenntnis* (Nuremberg, 1919)

H. Rutters: *Gustav Mahler* (Baarn, 1919; Eng. trans., 1953)

C. vom Wessem: *Gustav Mahler* (Arnhem, 1920)

R. Mengelberg: *Gustav Mahler* (Leipzig, 1923)

W. Hutschenruyter: *Mahler* (The Hague, 1927)

H. Holländer: 'Gustav Mahler', *MQ*, xvii (1931), 449

G. Engel: *Gustav Mahler: Song-Symphonist* (New York, 1932/*R*1970)

B. Walter: *Gustav Mahler* (Vienna, 1936, 2/1957; Eng. trans., 1937, 2/1941/*R*1970, with suppl. study by E. Krenek)

M. Carner: *Of Men and Music* (London, 1944)

W. Mellers: *Studies in Contemporary Music* (London, 1947)

D. Newlin: *Bruckner, Mahler, Schoenberg* (New York, 1947, rev. 2/1978) [with extensive bibliography]

A. Mathis: 'Gustav Mahler: Composer–Conductor', *The Listener*, xxxix (1948), 236

N. Loeser: *Gustav Mahler* (Antwerp, 1950)

W. Abendroth: *Vier Meister der Musik: Bruckner, Mahler, Reger, Pfitzner* (Munich, 1952)

E. Stein: *Orpheus in New Guises* (London, 1953)

H. F. Redlich: *Bruckner and Mahler* (London, 1955, rev. 2/1963)

A. Schibler: *Zum Werk Gustav Mahlers* (Lindau, 1955)

D. Mitchell: *Gustav Mahler: the Early Years* (London, 1958, rev. 2/1980)

T. W. Adorno: *Mahler: eine musikalische Physiognomik* (Frankfurt am Main, 1960, 5/1976)

D. Cooke: *Gustav Mahler: 1860–1911* (London, 1960, enlarged 2/1980)

F. Sopeña: *Introdución a Mahler* (Madrid, 1960)

T. Gedeon and M. Miklós: *Gustav Mahler* (Budapest, 1965)

T. W. Adorno and others: *Gustav Mahler* (Tübingen, 1966)

M. Vignal: *Mahler* (Paris, 1966)

H. Kralik: *Gustav Mahler*, ed. F. Heller (Vienna, 1968)

K. Blaukopf: *Gustav Mahler oder der Zeitgenosse der Zukunft* (Vienna, 1969; Eng. trans., 1973)

——: 'Auf neuen Spuren zu Gustav Mahler', *HiFi Stereophonie*, v (1971), 356

U. Duse: *Gustav Mahler* (Turin, 1973)

H.-L. de La Grange: *Mahler*, i (New York, 1973; Fr. trans., rev., 1979); ii (Paris, 1983) [with extensive bibliography]

M. Kennedy: *Mahler* (London, 1974/*R*1977)

J. Matter: *Connaissance de Mahler* (Lausanne, 1974)

B. W. Wessling: *Gustav Mahlers: ein prophetisches Leben* (Hamburg, 1974)

D. Mitchell: *Gustav Mahler: the Wunderhorn Years* (London, 1975; diss., U. of Southampton, 1977)

173

K. Rozenshil'd: *Gustav Maler* (Moscow, 1975)

C. Floros: *Gustav Mahler* (Wiesbaden, 1977)

F. Sopeña: *Estudios sobre Mahler* (Madrid, 1976)

E. Gartenberg: *Mahler: the Man and his Music* (New York, 1978)

V. Karbusicky: *Gustav Mahler und seine Umwelt* (Darmstadt, 1978)

P. Ruzicka, ed.: *Mahler: eine Herausforderung* (*ein Symposion*) (Wiesbaden, 1977)

Gustav Mahler Kolloquium Wien 1979

W. Szmolyon: 'Ein Mahler-Kolloquium in Wien', *ÖMz*, xxxiv (1979), 508

R. Werba: 'Mahlers Weg nach Wien', *ÖMz*, xxxiv (1979), 486

F. Willnauer: *Gustav Mahler und die Wiener Oper* (Vienna, 1979)

A. Bata and A. Gador: 'Tizenegy kiadatlan Mahler-level a Zenemuveszeti Foiskola koenyvtaraban', *Magyar zene*, xxi/1 (1980), 86

D. Cooke: *Gustav Mahler: an Introduction to his Music* (London, 1980)

W. Schreiber: 'Internationales Mahler-Symposium in Düsseldorf', *ÖMz*, xxxv (1980), 42

H. H. Eggebrecht: *Die Musik Gustav Mahlers* (Munich, 1982)

Q. Principe: *Mahler* (Milan, 1983)

SPECIAL BIOGRAPHICAL AND CHARACTER STUDIES

R. Mengelberg, ed.: *Das Mahler-Fest: Amsterdam, Mai 1920: Vorträge und Berichte* (Vienna, 1920)

A. Rosenzweig: 'Wie Gustav Mahler seine "Achte" plante: Die erste handschriftliche Skizze', *Der Wiener Tag*, no.3607 (Vienna, 4 June 1933)

M. Carner: 'Gustav Mahler's Visit to London', *MT*, lxxvii (1936), 408; repr. in *Of Men and Music* (London, 1944)

A. Jemnitz: 'Gustav Mahler als königlicher ungarischer Hofoperndirektor', *Der Auftakt*, xvi (1936), 7, 63, 183

K. Blessinger: *Mendelssohn, Meyerbeer, Mahler: drei Kapitel Judentum in der Musik als Schlüssel zur Musikgeschichte des 19. Jahrhunderts* (Berlin, 1939)

T. Reik: *The Haunting Melody* (New York, 1953)

E. Stein: 'Mahler and the Vienna Opera', *Opera*, iv (1953), 4, 145, 200, 281

K. P. Bernet-Kempers: 'Mahler und Willem Mengelberg', *Kongressbericht: Wien Mozartjahr 1956*, 41

L. Kitzwegerer: *Alfred Roller als Bühnenbildner* (diss., U. of Vienna, 1959)

H. J. Schaefer: 'Gustav Mahlers Wirken in Kassel', *Musica*, xiv (1960), 350

D. Mitchell: 'Gustav Mahler: Prospect and Retrospect', *PRMA*, lxxxvii (1960–61), 83

Bibliography

M. Brod: *Gustav Mahler: Beispiel einer deutsch-jüdischen Symbiose* (Frankfurt am Main, 1961)

D. Cvetko: 'Gustav Mahlers Saison 1881–82 in Laibach (Slovenien)', *Musik des Ostens*, iv (1968), 74

D. Kučerová: 'Gustav Mahler v Olomouci' [Mahler in Olomouc], *Časopis Vlastivědné společnosti muzejní v Olomouci* (1970), no.2–3, p.101

E. R. Reilly: 'Mahler and Guido Adler', *MQ*, lviii (1972), 436–70

M. L. von Deck: *Gustav Mahler in New York: his Conducting Activities in New York City, 1908–11* (diss., New York U., 1973)

W. J. McGrath: *Dionysian Art and Populist Politics in Austria* (New Haven, Conn., 1974)

H. H. Stuckenschmidt: *Schoenberg: Leben, Umwelt, Werk* (Zurich, 1974)

D. Holbrook: *Gustav Mahler and the Courage To Be* (London, 1975)

A. D. Keener: 'Gustav Mahler as Conductor', *ML*, lvi (1975), 341

V. Lébl: 'Gustav Mahler als Kapellmeister des deutschen Landestheaters in Prag', *HV*, xii (1975), 351 [incl. list of repertory for 1885–6 season]

R. Werba: 'Il "Don Giovanni" nella interpretazione di Gustav Mahler', *NRMI*, ix (1975), 515

S. Wiesmann, ed.: *Gustav Mahler und Wien* (Stuttgart and Zurich, 1976; Eng. trans., 1976)

K. Blaukopf: 'Gustav Mahler und Russland', *Neue Zürcher Zeitung* (1977), no.94, p.61

O. Brusatti: 'Mahler dirigiert Mahler: die Interpretations vorbereitungen von "Das Lied von der Erde" ', *Protokolle*, lxxvii (1977)

J. Matter: 'Mahler et Adorno', *Revue musicale de Suisse Romande*, xxx (1977), 156

S. Feder: 'Gustav Mahler, Dying', *International Review of Psychoanalysis*, v (1978), 125

C. Matthews: *Mahler at Work: Aspects of the Compositional Process* (diss., U. of Sussex, 1978)

R. P. Morgan: 'Ives and Mahler: Mutual Responses at the End of an Era', *19th Century Music*, ii (1978–9), 72

E. R. Reilly: *Gustav Mahler und Guido Adler: zur Geschichte einer Freundschaft* (Vienna, 1978; Eng. trans., 1982)

P. Banks: *The Early Social and Musical Environment of Gustav Mahler* (diss., U. of Oxford, 1980)

Gustav Mahler in Toblach (Vienna, 1981)

S. Feder: 'Gustav Mahler: the Music of Fratricide', *International Review of Psycho-analysis*, viii (1981), 1

P. Kuret: 'Gustav Mahler in Anton Krisper', *MZ*, xvii (1981), 77 [incl. Eng. summary]

L. Pinzauti: 'Mahler e Puccini', *NRMI*, xvi (1982), 330

H. J. Schaefer: *Gustav Mahler in Kassel* (Kassel, 1982)

MUSICAL STYLE

E. Stein: 'Mahlers Instrumentationsretuschen', *Pult und Taktstock*, iv (1928) and *Musikblätter des Anbruch*, x (1928), 42; Eng. trans. in *Orpheus in New Guises* (London, 1953)

E. Wellesz: *Die neue Instrumentation* (Berlin, 1928, 2/1929)

S. Langford: *Music Criticisms*, ed. N. Cardus (London, 1929)

E. Stein: 'Mahlers Sachlichkeit', *Musikblätter des Anbruch*, xii (1930), 99; Eng. trans. in *Orpheus in New Guises* (London, 1953)

A. Schaefers: *Gustav Mahlers Instrumentation* (Düsseldorf, 1935)

H. Tischler: *Die Harmonik in den Werken Gustav Mahlers* (diss., U. of Vienna, 1937)

———: 'Musical Form in Gustav Mahler's Works', *Musicology*, ii (1949), 231

A. Schoenberg: 'Gustav Mahler', *Style and Idea* (New York, 1950, enlarged, 1972, by L. Stein)

H. Tischler: 'Mahler's Impact on the Crisis of Tonality', *MR*, xii (1951), 113

H. Truscott: 'Some Aspects of Mahler's Tonality', *MMR*, lxxxvii (1957), 203

C. Matthews: 'Mahler at Work: some Observations on the Ninth and Tenth Symphony Sketches', *Soundings*, iv (1974), 76

A. Forchert: 'Zur Auflösung traditioneller Formkatagorien in der Musik um 1900: Probleme formaler Organisation bei Mahler und Strauss', *AMw*, xxxii (1975), 85

R. McGuinness: 'Mahler und Brahms: Gedanken zu "Reminiszenzen" in Mahlers Sinfonien', *Melos/NZM*, iii (1977), 215

M. Zenek: 'Die Aktualität Gustav Mahlers als Problem der Rezeptionsästhetik: Perspektiven von Mahlers Naturerfahrung und Formen ihrer Rezeption', *Melos/NZM*, iii (1977), 225

P. Chamonard: *Caractère et évolution de l'orchestre dans l'oeuvre symphonique de G. Mahler* (diss., U. of Paris, 1978)

K. Velten: 'Über das Verhältnis von Ausdruck und Form im Werk Gustav Mahlers und Anton von Weberns', *Musik und Bildung*, x (1978), 159

S. Vill: *Vermittlungsformen verbalisierter und musikalischer Inhalte in der Musik Gustav Mahlers* (Tutzing, 1979)

K. H. Stahmer and others, eds.: *Form und Idee in Gustav Mahlers Instrumentalmusik* (Wilhelmshaven, 1980)

M. Zenck: 'Mahlers Streichung des Waldmärchens aus dem Klagenden Lied; zum Verhältnis von philologischer Erkenntnis und Interpretation', *AMw*, xxxviii (1981), 179

J. van Holen: 'G. Mahler als orkestrator: nieuwe perspectieven',

Bibliography

Adem, xviii (1982), 226 [incl. Eng. and Fr. summaries]

D. Cooke: 'Mahler's Melodic Thinking', *Vindications: Essays on Romantic Music*, ed. D. Matthews (London, 1982)

T. Schmitt: *Der langsame Symphoniesatz Gustav Mahlers: historisch-vergleichende Studien zu Mahlers Kompositionstechnik* (Munich, 1983)

SYMPHONIES AND SONGS: GENERAL

E. Istel, ed.: *Mahlers Symphonien* (Berlin, *c*1910, 2/1920)

P. Bekker: *Gustav Mahlers Sinfonien* (Berlin, 1921/*R*1969)

F. E. Pamer: *Gustav Mahlers Lieder: eine Stilkritische Studie* (diss., U. of Vienna, 1922); abridged in *SMw*, xviii (1929), 116; xix (1930), 105

K. S. Sorabji: 'Notes on the Symphonies of Mahler', *Around Music* (London, 1932)

E. Wellesz: 'The Symphonies of Gustav Mahler', *MR*, i (1940), 2

S. Vestdijk: *Gustav Mahler: over de structuur van zijn symfonisch oeuvre* (The Hague, 1960)

N. Cardus: *Gustav Mahler: his Mind and his Music*, i (London, 1965)

T. W. Adorno: 'Zu einer imaginären Auswahl von Liedern Gustav Mahlers', *Impromptus* (Frankfurt am Main, 1968), 30

P. Barford: *Mahler: Symphonies and Songs* (London, 1970)

Z. Roman: *Mahler's Songs and their Influence on his Symphonic Thought* (diss., U. of Toronto, 1970)

M. Tibbe: *Über die Verwendung von Liedern und Liedelementen in instrumentalen Symphoniesätzen Gustav Mahlers* (Munich, 1971, rev. 2/1977)

E. F. Kravitt: 'Tempo as an Expressive Element in the Late Romantic Lied', *MQ*, lix (1973), 497

Z. Roman: 'Structure as a Factor in the Genesis of Mahler's Songs', *MR*, xxxv (1974), 157

I. Barsova: *Simfonii Gustava Malera* (Moscow, 1975)

E. W. Murphy: 'Sonata-rondo Form in the Symphonies of Gustav Mahler', *MR*, xxxvi (1975), 54

J. G. Williamson: *The Development of Mahler's Symphonic Technique with Special Reference to the Compositions of the Period 1899 to 1905* (diss., U. of Liverpool, 1975)

W. Berny-Negrey: 'Architektonika symfonii Gustawa Mahlera', *Muzyka*, xxii/4 (1977), 38 [incl. Eng. summary]

O. Kolleritsch, ed.: *Gustav Mahler: Sinfonie und Wirklichkeit*, Studien zur Wertungsforschung, ix (Graz, 1977)

D. de la Motte: 'Das komplizierte Einfache – zum ersten Satz der 9. Sinfonie von Gustav Mahler', *Musik und Bildung*, x (1978), 145

K. von Fischer: 'Bemerkungen zu Gustav Mahlers Liedern', *MZ*, xiii (1977), 57

——: 'Gustav Mahlers Umgang mit Wunderhorntexten', *Melos/NZM*, iv (1978), 103

B. Sponheuer: *Logik des Zerfalls: Untersuchungen zum Finalproblem in den Symphonien Gustav Mahlers* (Tutzing, 1978)

E. M. Dargie: *Music and Poetry in the Songs of Gustav Mahler* (diss., U. of Aberdeen, 1979)

P. Revers: *Die Liquidation der musikalischen Struktur in den späten Symphonien Gustav Mahlers* (diss., U. of Salzburg, 1980)

E. M. Dargie: *Music and Poetry in the Songs of Gustav Mahler* (Berne, 1981)

H. H. Eggebrecht: 'Symphonische Dichtung', *AMw*, xxxix (1982), 221ff

SYMPHONIES AND SONGS: INDIVIDUAL WORKS

H. Jalowetz: 'Mahler über die achte Symphonie', *Musikblätter des Anbruch*, v (1923), 135

R. Specht: Preface to *Gustav Mahler: nachgelassene Zehnte Symphonie* (Vienna, 1924) [facs. edn.]

E. Stein: 'Die Tempogestaltung in Mahlers neunter Sinfonie', *Pult und Taktstock* (1924), Oct–Nov; Eng. trans. in *Orpheus in New Guises* (London, 1953)

——: 'Eine unbekannte Ausgabe letzter Hand von Mahlers 4. Sinfonie', *Pult und Taktstock* (1929), March–April; Eng. trans. in *Orpheus in New Guises* (London, 1953)

H. Boys: 'Mahler and his Ninth Symphony' (1938) [HMV disc notes]

M. Carner: 'Form and Technique in Mahler's "Lied von der Erde" ', *MMR*, lxix (1939), 48; repr. in *Of Men and Music* (London, 1944)

A. Mathis: 'Mahler's Unfinished Symphony', *The Listener*, xl (1948), 740

H. Tischler: 'Mahler's "Das Lied von der Erde" ', *MR*, x (1949), 111

E. W. Mulder: *Das Lied von der Erde: een critisch-analytische studie* (Amsterdam, 1951)

E. Ratz: 'Zum Formproblem bei Gustav Mahler: eine Analyse des ersten Satzes der IX. Sinfonie', *Mf*, viii (1955), 169; pubd separately in T.W. Adorno and others: *Gustav Mahler* (Tübingen, 1966), 123

——: 'Zum Formproblem bei Gustav Mahler: eine Analyse des Finales der VI. Sinfonie', *Mf*, ix (1956), 156; pubd separately in T. W. Adorno and others: *Gustav Mahler* (Tübingen, 1966), 90; Eng. trans., *MR*, xxix (1968), 34

A. Berg: 'Über Mahlers IX. Symphonie', H. F. Redlich: *Alban Berg: Versuch einer Würdigung* (Vienna, 1957)

J. Diether: 'The Expressive Content of Mahler's Ninth', *Chord and Discord*, ii/10 (1963), 69

D. Mitchell: 'Mahler's Enigmatic Seventh Symphony', *The Listener*, lxix (1963), 649

Bibliography

H. F. Redlich: 'Mahler's Enigmatic "Sixth"', *Festschrift Otto Erich Deutsch* (Kassel, 1963), 250

E. Ratz: Preface to *Gustav Mahler: IX Symphonie* (Vienna, 1971) [facs. edn.]

J. L. Broeckx: *Gustav Mahler's Das Lied von der Erde* (Antwerp, 1975)

K. von Fischer: 'Die Doppelschlagfigur in den zwei letzten Sätzen von Gustav Mahlers 9. Symphonie', *AMw*, xxxii (1975), 99

Z. Roman: 'Aesthetic Symbiosis and Structural Metaphor in Mahler's *Das Lied von der Erde*', *Festschrift Kurt Blaukopf* (Vienna, 1975), 110

P. Andraschke: *Gustav Mahlers IX. Symphonie: Kompositionsprozess und Analyse* (Wiesbaden, 1976)

D. Cooke: 'The History of Mahler's Tenth Symphony', *Gustav Mahler: a Performing Version of the Draft for the Tenth Symphony* (London and New York, 1976); repr. as 'Mahler's Tenth Symphony', *MT*, cxvii (1976), 563, 645

C. Matthews and D. Mitchell, eds.: *G. Mahler: Lieder eines fahrenden Gesellen* (London, 1977) [vocal score with critical introduction and notes]

S. M. Filler: *Editorial Problems in Symphonies of Gustav Mahler: a Study of the Sources of the Third and Tenth Symphonies* (diss., Northwestern U., Ill., 1977)

P. R. Franklin: 'The Gestation of Mahler's Third Symphony', *ML*, lviii (1977), 439

A. Wenk: 'The Composer as Poet in *Das Lied von der Erde*', *19th Century Music*, i (1977–8), 33

P. Andraschke: 'Struktur und Gehalt im ersten Satz von Gustav Mahlers Sechster Symphonie', *AMw*, xxxv (1978), 275

R. A. Kaplan: 'Interpreting Surface Harmonic Connections in the Adagio of Mahler's Tenth Symphony', *In Theory Only*, iv/2 (1978), 32

E. Klemm: 'Gustav Mahlers X. Sinfonie', *Musik und Gesellschaft*, xxviii (1978), 549

E. F. Kravitt: 'Mahler's Dirges for his Death: February 24, 1901', *MQ*, lxiv (1978), 329 [*Kindertotenlieder*]

P. Revers: 'Liquidation als Formprinzip: die formprägende Bedeutung des Rhythmus für das Adagio der 9. Symphonie von Gustav Mahler', *ÖMz*, xxxiii (1978), 527

H. P. Rosack: *Symphony no.10 by Gustav Mahler: the Fundamental Performance Problems of the Adagio* (diss., Stanford U., 1979)

R. Stephan: *Gustav Mahler, II. Symphonie C-moll*, Meisterwerke der Musik, xxi (Munich, 1979)

P. Bergquist: 'The First Movement of Mahler's Tenth Symphony: an Analysis and an Examination of the Sketches', *Music Forum*, v

(1980), 335–94

N. Del Mar: *Mahler's Sixth Symphony: a Study* (London, 1980)

S. Feder: 'Gustav Mahler um Mitternacht', *International Review of Psycho-analysis*, vii (1980), 11

R. F. Jones: *Thematic Development and Form in the First and Fourth Movements of Mahler's First Symphony* (diss., Brandeis U., 1980)

V. Kalisch: 'Bemerkungen zu Gustav Mahlers Kindertotenlieder: dargestellt am Beispiel des zweiten', *MZ*, xvi (1980), 31

J. Williamson: 'Mahler's Compositional Process: Reflections on an Early Sketch for the Third Symphony's First Movement', *ML*, lxi (1980), 338

S. M. Filler: 'The Case for a Performing Edition of Mahler's Tenth Symphony', *Journal of Musicological Research*, iii (1981), 274

R. A. Kaplan: 'The Interaction of Diatonic Collections in the Adagio of Mahler's Tenth Symphony', *In Theory Only*, vi/1 (1981), 29

D. Cooke: 'The Facts Concerning Mahler's Tenth Symphony'; 'The Word and the Deed: Mahler and his Eighth Symphony', *Vindications: Essays on Romantic Music*, ed. D. Matthews (London, 1982)

R. Gerlach: *Strophen von Leben, Traum und Tod: ein Essay über die Rückert-Lieder von Gustav Mahler* (Wilhelmshaven, 1982)

R. Stephan: 'Überlungen zur Taktgruppenanalyse: zur Interpretation der 7. Symphonie von Gustav Mahler', *Logos musicae: Festschrift für Albert Palm* (Wiesbaden, 1982)

J. Williamson: 'Deceptive Cadences in the Last Movement of Mahler's Seventh Symphony', *Soundings*, ix (1982), 87

C. M. Zenck: 'Zur Vorgeschichte der Uraufführung von Mahlers Zehnter Symphonie', *AMw*, xxxix (1982), 245

H. Danuser: 'Gustav Mahlers Symphonie "Das Lied von der Erde" als Problem der Gattungsgeschichte', *AMw*, xl (1983), 276

V. Kofi Agawu: 'The Musical Language of Kindertotenlieder No.2', *Journal of Musicology*, ii (1983), 81

JUVENILIA

H. Holländer: 'Ein unbekannter Teil von Gustav Mahlers "Klagendem Lied" ', *Der Auftakt*, xiv (1934), 200

——: 'Mahler-Uraufführung in Brunn', *Musikblätter des Anbruch*, xvi (1934), 201 [*Waldmärchen*]

D. Newlin: 'Gustav Mahler's Piano Quartet in A minor (1876)', *Chord and Discord*, ii/10 (1963), 180

J. Diether: 'Mahler's "Klagende Lied": Genesis and Evolution', *MR*, xxix (1968), 268

——: 'Notes on Some Mahler Juvenilia', *Chord and Discord*, iii/1 (1969), 3–100

D. Mitchell: 'Mahler's Waldmärchen', *MT*, cxi (1970), 375

P. Boulez: Preface to *Gustav Mahler: Das klagende Lied* (Vienna and

Bibliography

London, 1971) [Philharmonia Pocket Score no.392]

D. Newlin: 'Mahler's Opera', *Opera News*, xxxvi (1972), 6 [*Rübezahl*]

P. Banks: 'An Early Symphonic Prelude by Mahler?', *19th Century Music*, iii (1979–80), 141

H. Danuser: 'Mahlers Lied "Von der Jugend": ein musikalisches Bild', *Art nouveau – Jugendstil und Musik* [Festschrift for Willi Schuh's 80th birthday] (Zurich, 1980)

S. E. Hefling: 'The Road not Taken: Mahler's Rübezahl', *Yale University Library Gazette*, lvii (1983), 145

PERFORMING EDITIONS

G. Adler: ' "Euryanthe" in neuer Einrichtung', *ZIMG*, v (1903–4), 267

G. Brecher: *Oberon, König der Elfen: neue Bühneneinrichtung von Gustav Mahler* (Vienna, 1914) [incl. Mahler's lib trans.]

L. Hartmann: *Opernführer: Die drei Pintos* (Leipzig, n.d.)

M. Carner: 'Mahler's Re-scoring of the Schumann Symphonies', *MR*, ii (1941), 97; repr. in *Of Men and Music* (London, 1944)

J. Warrack: *Carl Maria von Weber* (London, 1968, 2/1976) [incl. chap. on Mahler's reconstruction of *Die drei Pintos*]

P. Andraschke: 'Die Retuschen Gustav Mahlers an der 7. Symphonie von Franz Schubert', *AMw*, xxxii (1975), 106

E. Hilmar: ' "Schade, aber es muss(te) sein": zu Gustav Mahlers Strichen und Retuschen insbesondere am Beispiel der V. Symphonie Anton Bruckners', *Bruckner-Studien*, ed. O. Wessely (Vienna, 1975), 187

H. Blaukopf: 'Eine Oper "aus Weber" ', *ÖMz*, xxxiii (1978), 204 [*Die drei Pintos*]

V. Kalisch: 'Zu Mahlers Instrumentationsretuschen in den Sinfonien Beethovens', *SMz*, cxxi (1981), 17

D. McCaldin: 'Mahler and Beethoven's Ninth Symphony', *PRMA*, cvii (1981–2), 101

181

RICHARD STRAUSS

Michael Kennedy

Robert Bailey

CHAPTER ONE

Boyhood and youth, 1864–85

Richard (Georg) Strauss, born in Munich on 11 June 1864, was the first child of Franz Joseph Strauss (1822–1905), principal horn in the Munich Court Orchestra for 42 years, and Josephine Pschorr (1837–1910), his second wife. Fräulein Pschorr was a member of the family of brewers, which enabled her husband to enjoy financial independence and meant that Strauss and his sister had a happy, carefree childhood. Strauss showed musical promise from his earliest years: at the age of four he had piano lessons from his father's orchestral colleague August Tombo, and four years later he was taught the violin by his father's cousin Benno Walter, leader of the court orchestra. Franz Strauss was intensely conservative in his musical tastes – though he played Wagner magnificently he detested both the music and the man – and did not allow his son to hear anything but the classics until he was in his early teens. A powerful influence on the boy was his freedom to attend rehearsals of the Munich Court Orchestra under Hermann Levi. From the age of 11 he received instruction in theory, harmony and instrumentation from one of Levi's assistant conductors, Friedrich Wilhelm Meyer. His first compositions were written when he was six, and from then until the last days of his life he composed regularly and copiously.

After elementary schooling Strauss entered the

Ludwigsgymnasium, Munich, in 1874, remaining there until he matriculated at the age of 18. (He never went to an academy of music, going from school to Munich University for the winter and spring terms of 1882–3 to read philosophy, aesthetics and the history of art.) His first encounter with the operas of Wagner came during and after 1874, when he saw performances of *Tannhäuser*, *Siegfried* and *Lohengrin* and found them beyond his power of appreciation. 'It was not until, against my father's orders, I studied the score of *Tristan*, that I entered into this magic work, and later into *Der Ring des Nibelungen*, and I can well remember how, at the age of seventeen, I positively wolfed the score of *Tristan* as if in a trance', he wrote many years later in his reminiscences (1949). Some years earlier Franz Strauss had formed a semi-professional orchestra known as 'Wilde Gung'l'. From the age of 13 Strauss was allowed to play at a back desk of the violins and he gradually moved up to the front desks. A particularly significant month was March 1881, when he was 16. On the 14th his String Quartet in A was performed in Munich by Benno Walter's quartet; on the 26th the Wilde Gung'l performed his *Festmarsch* in E♭ and on the 30th Levi conducted the court orchestra in the Symphony in D minor. That year he composed the Piano Sonata in B minor op.5 and the Five Piano Pieces op.3. All these juvenilia were published and are extant, with the exception of the symphony, which survives in manuscript.

Strauss's career as a composer began in earnest when performances of his works were given outside his native Munich where he enjoyed a favoured position. On 5 December 1882 his Violin Concerto was performed in

Vienna by Benno Walter with Strauss as pianist. But of more importance to his future was the performance a week earlier by the Dresden Court Orchestra under Franz Wüllner of the Serenade in E♭ for 13 wind instruments. In winter 1883–4, by which time he had left university in order to concentrate on music, Strauss visited Berlin for the first time. During his stay he heard operas, met the city's artistic circle, developed his lifelong addiction to card playing and wrote another symphony. He also met Hans von Bülow, at this time conductor of the Meiningen Court Orchestra, which he had made into the most disciplined ensemble in Europe. Strauss's first publisher, Eugen Spitzweg, knew Bülow and had sent him the score of Strauss's Serenade. Bülow was impressed enough not only to put the work into his orchestra's repertory but to describe Strauss as 'by far the most striking personality since Brahms'. The Serenade was performed in Berlin, in Strauss's presence, by the Meiningen Orchestra, after which Bülow invited the 19-year-old composer to write a similar piece for Meiningen, the result being the Suite in B♭. Bülow arranged for the first performance to be given in Munich when the Meiningen Orchestra played there on 18 November 1884 and invited Strauss to conduct his piece.

The years 1881–5 were highly productive: Strauss composed the Horn Concerto no.1, the Cello Sonata, the *Stimmungsbilder* for piano, the Piano Quartet, the Goethe setting *Wandrers Sturmlied* for six-part chorus and orchestra, the Symphony no.2 in F minor and nine settings of poems by Gilm for voice and piano which include *Zueignung*, *Die Nacht* and *Allerseelen* (still among the most admired of Strauss lieder). The symphony had its first performance in New York, conducted by the enterprising Theodore Thomas, who when visiting Europe had been shown the manuscript by Franz Strauss. Early in 1885 Wüllner conducted this symphony in Cologne; and at Meiningen on 4 March Bülow conducted the first performance of the Horn Concerto no.1. Before his 21st birthday, therefore, Strauss had heard his music interpreted by the outstanding German conductors of the day, one of whom, Bülow, had dubbed him 'Richard the Third', a jest because he meant that after 'Richard the First' (Wagner) there could be no Richard the Second. It certainly could not have meant that Bülow regarded Strauss as a revolutionary innovator, because what he admired in the young composer's music was its adherence to traditional practices. Nevertheless it was an astonishing compliment, and indicative of Strauss's rapid rise to fame.

CHAPTER TWO

Conductor and tone-poet, 1885–98

In the summer of 1885, when the position of assistant conductor to Bülow at Meiningen became vacant, Bülow offered it to Strauss who, in spite of his inexperience with the baton, accepted. He attended all Bülow's rehearsals, learning by watching and by answering Bülow's searching questions about scores. On 15 October, after only a fortnight, Strauss made his public début as solo pianist in Mozart's C minor Concerto K491, for which he composed cadenzas, and conducted his own F minor Symphony. Brahms was in Meiningen for rehearsals of his new Symphony no.4, which was to have its first performance ten days later, and he listened to the young man's work, remarking that it was 'quite attractive' but 'too full of thematic irrelevances' (see Strauss's memoirs). A few days later Bülow resigned his post and Strauss was appointed his successor by the Duke of Saxe-Meiningen. But without the kudos of Bülow as his conductor, the duke began to reduce the orchestra and Strauss, despite misgivings, accepted a three-year contract as third conductor at the Munich Court Opera. He left Meiningen in April 1886 having gained invaluable experience on the rostrum. He had also been an assiduous attender of performances by the famous Meiningen Court Theatre, where his deep knowledge of dramatic and theatrical effectiveness developed.

16. Richard Strauss, c1890

Another profound personal influence on Strauss at
Meiningen had been his friendship with one of the
orchestral violinists, Alexander Ritter, a devout follower
of Wagner and Liszt who had married Wagner's niece.
Ritter had found the young conductor–composer fertile
ground for conversion to the faith of 'Zukunftsmusik'
(music of the future). He had interested him in Wagner's
essays and in Schopenhauer and had persuaded him that
'new ideas must search for new forms – this basic
principle of Liszt's symphonic works, in which the
poetic idea was really the formative element, became
henceforward the guiding principle for my own symphonic work' (Strauss's memoirs). The immediate
musical result of this conversion was the symphonic
fantasy *Aus Italien*, a halfway stage between the
Mendelssohnian conventionality of his early works and
the Lisztian models to follow. It recorded the impres-
sions of his first visit to Italy in summer 1886, before he
took up his Munich post. He remained in Munich for
three years, somewhat fretfully because, as third conductor, he was denied the chance to direct the important
works. But among the lesser operas that came his way
were Mozart's *Così fan tutte* and Verdi's *Un ballo in
maschera*, a classification indicative of public taste at
the time. *Aus Italien* was first performed in Munich in
March 1887 and divided the audience into applauders
and booers, thereby giving this eminently uncontrover-
sial work a *cachet de scandale* which assisted its
progress. In 1887, while guest conducting in Leipzig,
Strauss met Mahler, whom he immediately admired and
liked; and on his summer holiday that year he was asked
to give some singing lessons to a young soprano, Pauline
de Ahna, who was the daughter of a Wagner-loving

general. The principal work on which he was engaged was a symphonic poem – or tone poem, to use the term he preferred – based on Shakespeare's *Macbeth*, and when it was completed in 1888 he at once began another, *Don Juan*. His mentor was still Bülow, who suggested revisions in *Macbeth* and also recommended him to the Weimar Opera as assistant conductor.

Strauss left Munich in 1889; during that summer he worked as répétiteur at Bayreuth, where he found favour with Cosima Wagner. He took with him to Weimar the completed score of *Don Juan* and the two tasks on which he was working, his libretto for an opera, *Guntram*, and a rough sketch of a tone poem, *Tod und Verklärung*. His employers at Weimar were deeply impressed by *Don Juan* when he played it to them on the piano, and they insisted that its first performance should be at a Weimar concert. Although he was doubtful of the orchestra's ability to cope with the work's unprecedented technical difficulties, Strauss conducted it on 11 November 1889. It was his biggest triumph to date and thenceforward he was generally regarded as the most significant and progressive German composer since Wagner. Seven months later, at an Eisenach concert, he conducted the first performances of his *Burleske* for piano and orchestra and *Tod und Verklärung*. The following October he conducted the revised *Macbeth* at Weimar. Meanwhile Pauline de Ahna had joined the Weimar company and sang Isolde in *Tristan* when Strauss conducted it, uncut, after advice from Cosima, in January 1892. With *Così fan tutte* it remained Strauss's favourite opera throughout his life.

In June 1892 Strauss was seriously ill and spent the winter in Egypt. He completed the music of *Guntram* in

Cairo and conducted the first performance in Weimar in May 1894 with Pauline as the heroine Freihild. An indication of Strauss's reputation at this date is that, after Bülow's death in 1894, the Berlin PO invited him to take over Bülow's concerts, but the venture was not a success, Strauss admitting that he was not yet ready for such a post. He had, however, been asked back to Munich as associate to the ailing Levi and had accepted, though with reluctance, on Cosima Wagner's advice. He was to begin his duties there on 1 October 1894; in the preceding August he conducted for the first time at Bayreuth (*Tannhäuser*) and on 10 September he married Pauline. His wedding present to her was the four superb songs of his op.27, *Morgen, Cäcilie, Ruhe, meine Seele* and *Heimliche Aufforderung*.

With Levi frequently ill, Strauss had the satisfaction in Munich of conducting *Tristan* and *Die Meistersinger* and a Mozart festival of *Die Entführung, Così fan tutte* and *Don Giovanni*. In his second season *Guntram* was staged for one disastrous performance. Two of the leading singers refused to take part and the orchestra petitioned the Intendant to spare them from 'this scourge of God'. Strauss was embittered by this attitude in his native city and eventually had his revenge. The years 1894–9 were exceptionally prolific in compositions. In addition to many lieder he wrote, one after the other, four of his best orchestral works, *Till Eulenspiegels lustige Streiche* (1894–5), *Also sprach Zarathustra* (1895–6), *Don Quixote* (1896–7) and *Ein Heldenleben* (1897–8). All were well received and consolidated his position as the outstanding composer of his day, regarded as the arch-fiend of modernism and cacophony because of the huge instrumental forces, the innovatory design and the naturalistic effects he employed.

But it is easy to overlook Strauss's ability and reputation as a conductor, which he regarded as his principal role at this time. The musical life of Germany and Austria in the 1890s and 1900s was dominated by three conductors who were known also as composers, Mahler, Strauss and Weingartner. Strauss was in constant demand as guest conductor of his own works and he visited Holland, Spain, France and England in 1897, the year in which his son was born. By now he was chief conductor of the Munich Opera, Levi having retired in 1896, but he had no hesitation in 1898 in accepting the post of chief conductor of the Royal Court Opera in Berlin. This was no sinecure: in his first eight-month season he conducted 71 performances of 25 operas, including two first performances and a *Ring* cycle. In the ensuing years he made many conducting tours and also conducted concerts of the Berlin Tonkünstler Orchestra and later the Berlin PO. He was ever ready to champion the unfamiliar and new. At Weimar he had revived several Liszt works (such as the *Faust Symphony*) and these remained in his repertory. In Berlin he conducted the music of Reger, Mahler, Schillings, Sibelius and Elgar. Another important feature of Strauss's life at this period was the beginning in 1898 of his seven-year campaign for a revision of German copyright law and the establishment of a performing-right society. He was always alive to the value of money, and many are the gibes about his constant talk of royalties. But he saw no reason why a composer should not be well remunerated for his work and persistently championed his colleagues' rights in this respect as well as his own.

194

CHAPTER THREE

The opera composer, 1898–1918

In 1898 Strauss met the poet and satirist Ernst von Wolzogen (1855–1934) to whom he confided his wish to 'wreak some vengeance' on Munich for the way it had treated him over *Guntram* by composing an opera lampooning the city's philistinism. Strauss found a suitable subject in a Flemish medieval legend. Wolzogen transferred the action to medieval Munich and introduced puns and allusions into the text about the sorcerer, Richard Wagner, and his apprentice Strauss, thus providing Strauss with an equal opportunity for musical jokes and quotations. The one-act opera was called *Feuersnot* and was completed in 1901. Its première was in Dresden, under Ernst von Schuch, thereby inaugurating a long association between Strauss and the Dresden Opera. *Feuersnot* was an instant success, being a considerable advance on the pseudo-Wagnerian *Guntram*, and it was introduced to Vienna by Mahler and to London (in 1910) by Beecham. In 1903 Strauss completed another large-scale orchestral work, the *Symphonia domestica*, which described events in his home life. He conducted the first performance on 21 March 1904 in New York on his first visit to the USA. The content of the work caused a furore, but this was nothing compared with the sensational reaction to his next opera, *Salome*, a setting of a German transla-

195

17. Design by Alfred Roller for Act 3 of 'Der Rosenkavalier', first performed at the Dresden Court Opera on 26 January 1911

tion of Oscar Wilde's play. It was produced at Dresden on 9 December 1905. Almost everywhere *Salome* ran into censorship trouble and was regarded as the ultimate in salacious and blasphemous art. But this merely provided profitable publicity for the opera, which was a great success with the public, who like to be shocked, and was performed by 50 opera houses within two years. With the royalties Strauss built the villa at Garmisch in which Pauline and he lived from 1908 until the ends of their lives. Strauss followed *Salome* with *Elektra*, another one-act opera about an obsessed woman, his first collaboration with the Austrian poet and dramatist Hugo von Hofmannsthal. Strauss had seen Max Reinhardt's stage production of Hofmannsthal's version of Sophocles and had asked the poet if he would adapt it as a libretto. From the first Strauss recognized the possibilities in a permanent association with Hofmannsthal, poles apart though they were in character and outlook. 'We were born for one another and are certain to do fine things together', he wrote. *Elektra* was produced at Dresden under Schuch on 25 January 1909. It failed to make as great an impression as *Salome* had, but opera houses were eager to stage it and it enjoyed notoriety as the height (or depth) of cacophonous modernity, even Ernest Newman referring to the music as 'abominably ugly'.

The next Strauss–Hofmannsthal collaboration was a three-act 'comedy for music', *Der Rosenkavalier*, set in the 18th-century Vienna of the Empress Maria Theresia and making glorious anachronistic use of the waltz. It was composed between spring 1909 and September 1910. Schuch again conducted the first performance, lavishly produced under Reinhardt's supervision and with settings by Alfred Roller, at Dresden on 26 January 1911. By now Strauss's operas were awaited with the highest expectations and attended by intensive advance publicity. Special *Rosenkavalier* trains ran from Berlin to Dresden, and several other opera houses produced the work within days of the Dresden première. The opera won an immediate public acclaim which has never abated; indeed it has intensified. *Der Rosenkavalier* was followed in 1912 by the unusual experiment of the first version of *Ariadne auf Naxos*, in which the first part was a performance of Molière's *Le bourgeois gentilhomme* with incidental music by Strauss, and the second a one-act opera *Ariadne auf Naxos*, in which characters from the *commedia dell'arte* intermingled with the mythological figures. It was not a success and had little prospect of a future because such a hybrid depended on the extreme difficulty of obtaining first-rate companies of actors and singers simultaneously. In 1916 Strauss and Hofmannsthal, who had come near to a split over this work, revised it by scrapping the Molière play and substituting a sung prologue which proved to be one of Strauss's most effective and novel stage pieces. In this form the work was produced in Vienna on 4 October 1916, with Lotte Lehmann singing the travesty role of the Composer in the prologue, Maria Jeritza as Ariadne and Selma Kurz as Zerbinetta. Notwithstanding this vocal galaxy, it was still not a success and made its way comparatively slowly towards its present relative popularity. In fact, after *Der Rosenkavalier*, Strauss never again enjoyed unalloyed success with any of his works. Those who expected him to continue to provide the sensational frissons of *Salome* and *Elektra* were convinced that

18. *Playbill for the first performance of the one-act version of 'Ariadne auf Naxos', Stuttgart, 1912*

with *Rosenkavalier* he had 'gone soft', changed direction and retreated into comfortable note-spinning.

Strauss's first non-operatic collaboration with Hofmannsthal was in 1912–14 on a ballet for Dyagilev, *Josephs-Legende*, for which Hofmannsthal and Count Harry Kessler devised the scenario. Produced on an excessively lavish scale in Paris on 14 May 1914, it was also staged in London at Drury Lane on 23 June conducted by Beecham, who had introduced all Strauss's existing operas except *Guntram* to London between 1910 and 1913. While in England Strauss received the honorary degree of DMus from Oxford University to mark his 50th birthday. He was already at work on another opera, *Die Frau ohne Schatten*, a symbolic fairy tale that Hofmannsthal described as 'related to *Zauberflöte* as *Rosenkavalier* is to *Figaro*'. Progress was interrupted by the outbreak of war in August 1914 (one consequence of which was the sequestration of a large part of his savings, which he had banked in England with the German-born financier Sir Edgar Speyer). Strauss therefore completed scoring the last of his important orchestral tone poems, *Eine Alpensinfonie*, on which he had been working intermittently since 1911, and conducted the first performance on 28 October 1915 in Berlin. In 1917 he then added more incidental music for a further Hofmannsthal adaptation of *Le bourgeois gentilhomme*. Strauss told Hofmannsthal in 1916 that he wanted to compose 'an entirely modern, absolutely realistic domestic and character comedy', but the poet responded with distaste and recommended Hermann Bahr as a possible librettist. Strauss outlined his ideas for an opera based on a marital misunderstanding between Pauline and himself some years earlier.

Bahr sketched out a libretto; Strauss replied with some of his own suggestions which Bahr enjoyed so much that he withdrew, insisting that in this case the composer should be his own librettist: in July 1917 Strauss completed the libretto of *Intermezzo*, but he did not complete the music until 1923.

CHAPTER FOUR

After World War I, 1919–49

From 1910 Strauss was a guest conductor of the Berlin Opera but this association ended in 1918. He then signed a five-year contract with the Vienna Staatsoper, to run from 1919, as joint director with Franz Schalk. Just before this association began Schalk conducted the first performance of *Die Frau ohne Schatten* on 10 October 1919. The work's difficulties in the matter of staging proved a severe handicap to its success, despite a brilliant cast. Strauss and Schalk had at their command the services of such singers as Jeritza, Lehmann, Kurz, Elisabeth Schumann, Maria Olczewska, Leo Slezak, Alfred Piccaver, Richard Tauber, Karl Aagard-Oestvig, Alfred Jerger and Richard Mayr. Strauss not only conducted but in effect produced the great classical repertory in a memorable manner. He was incontestably a great conductor. The demonstrative gestures of his youth had given way to a laconic almost motionless style, but its effectiveness was undeniable. Erich Kleiber has described how Strauss, like Nikisch, could produce tremendous crescendos in the final pages of *Tristan* simply by slowly raising his left hand. In spite of these artistic successes, however, there was continual tension between Schalk and Strauss; moreover, the Austrian civil servants who now ran the Staatsoper considered that Strauss was financially extravagant (his reply was: 'I am here to lose money'). The inevitable result was his engineered resignation in 1924, although to mark his 60th birthday the city of Vienna had presented him with a plot of land in the Belvedere on which he built a splendid house. He therefore remained an influential figure in the city and two years later returned to the Staatsoper as a guest conductor.

Strauss's creative output slackened during his Vienna conductorship, which accounted for his slow progress with *Intermezzo*. His ballet *Schlagobers* (completed 1922, performed 1924) was a flop, and in 1924 he wrote the *Parergon zur Symphonia domestica* for piano and orchestra for the one-armed pianist Paul Wittgenstein. *Intermezzo* was produced at Dresden on 4 November 1924, conducted by Fritz Busch. The partnership with Hofmannsthal was resumed in 1923 with the two-act opera *Die ägyptische Helena*, which Busch also conducted at its Dresden première on 6 June 1928. It was a failure there and in Vienna and New York; it strengthened the general contemporary critical view that Strauss the composer was an extinct volcano, out of touch with the postwar world and having not only nothing new but nothing at all to say. Since 1922 Strauss had persistently asked Hofmannsthal for 'another *Rosenkavalier*, without its mistakes and longueurs'. Hofmannsthal believed in 1928 that he had found the answer in *Arabella*, which was also set in Vienna but otherwise bore little resemblance to the earlier work. He and Strauss were still revising the completed libretto when on 15 July 1929 Hofmannsthal died from a stroke. Strauss began to compose the music as a tribute, but after rapidly completing the first act he faltered and did not finish the rest until 1932.

In the meantime, in 1931, Strauss had found a new

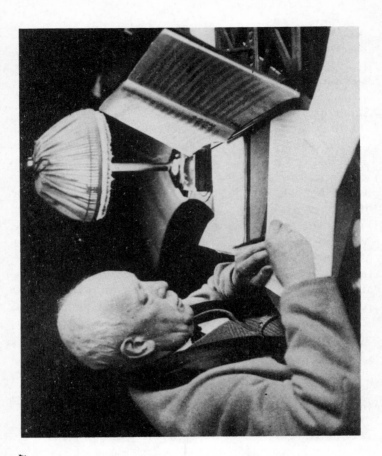

19. Strauss at work on 'Die schweigsame Frau'

librettist, the Jewish novelist and biographer Stefan Zweig, who offered him an adaptation of Ben Jonson's *Epicene, or The Silent Woman.* Strauss was delighted with it and in 1933 began to compose *Die schweigsame Frau.* But this was the year the National Socialists, with Hitler as chancellor, came to full power in Germany. Strauss had spent his life until he was over 50 as a court composer, accustomed to ignoring politics while he carried out his musical duties. He was totally obsessed by music, which was in every sense his life. He was contemptuous of most politicians: they came and went while he went his chosen way. Yet he was a German patriot too. Also, he was nearing 70, late in life to abandon one's home and country as other German musicians were doing. The Nazis realized that Strauss's eminence was valuable to them as propaganda. He played into their hands at the 1933 Bayreuth Festival when he conducted *Parsifal* in place of Toscanini, who had withdrawn in protest against the Nazis' attitude to the Jews. Strauss acted to save the festival and because of his veneration of Wagner, but his gesture was misunderstood by the opponents of Hitler and he was branded as a supporter of the dictator. Goebbels, minister of propaganda, capitalized on this in November 1933, when he established a state music bureau, the Reichsmusikkammer, and proclaimed Strauss as president without even consulting him.

The realities of the situation began to dawn on Strauss when German theatres were forbidden to produce works by Jews and he was denounced on the radio for working with Zweig. Wisely, Zweig went to Zurich and realized that he could no longer collaborate with Strauss. But Strauss could not accept this and suggested a secret arrangement. All that Zweig would agree to do was to suggest subjects and to supervise their compilation by others. Strauss replied to Zweig in an indiscreet letter posted in Dresden, where *Die schweigsame Frau* was in rehearsal, and which was intercepted by the Gestapo. He discovered that Zweig's name had been omitted from the posters and programmes and demanded its restoration. This was done, but Hitler and Goebbels, who had promised to attend the première, stayed away, and after four performances the opera was banned throughout Germany. At the same time Strauss was ordered to resign his presidency of the Reichs-musikkammer on the grounds of ill-health. Strauss then wrote an obsequious letter to Hitler, but he was now desperate to protect not himself but his daughter-in-law Alice, who was Jewish, and her children.

Zweig recommended the Viennese theatrical archivist Josef Gregor as librettist for the next opera subject that Strauss and he had selected, an episode in the Thirty Years War. Strauss did not much like Gregor or his work and treated him contemptuously. The opera was to be entitled *Friedenstag*, and Zweig revised the sections of libretto with which Strauss was dissatisfied. *Friedenstag* is a hymn to peace and it is extraordinary that it was accepted in Nazi Germany when it had its first performance in Munich in July 1938; indeed, it achieved 100 performances in two years. It is in one act and was intended as half of a double bill with *Daphne*, completed in 1937, though they are usually performed separately. Strauss's third opera with Gregor, *Die Liebe der Danae*, was a reworking of a scenario Hofmannsthal had sent to Strauss in 1920. As in *Daphne*, certain problems in the libretto were solved by the conductor

Clemens Krauss, who had been a favourite Strauss interpreter since the 1920s, and it was Krauss who became librettist of what proved to be Strauss's last opera, the 'conversation piece' *Capriccio*, composed in 1940–41 and first performed in Munich in October 1942.

After the outbreak of war in 1939 the 75-year-old Strauss was of little interest to the Nazi authorities, but in Garmisch his refusal to allow evacuees into his home resulted in ostracism for Alice Strauss and her children. In 1941 the Strauss family were permitted to reoccupy their Vienna home. Strauss was content anywhere, provided he could work, and in Vienna he wrote the Horn Concerto no.2 and the first of two sonatinas for wind instruments, almost, it would seem, in deliberate reversion to the type of music he wrote as a youth. As one might have expected, the enormities of what Nazi Germany had brought upon the world and itself only began to affect him when related to music. The destruction of the Munich Nationaltheater in 1943, followed by the destruction of the opera houses of Dresden, Weimar and Vienna, finally brought home to him the tragedy of his country. He had also fallen foul of Hitler's second-in-command, Martin Bormann, and was protected only by the favour of Baldur von Schirach, governor of Vienna. Although celebrations of Strauss's 80th birthday in June 1944 were officially discouraged, both Schirach and Krauss ensured that it was substantially marked. Strauss had adamantly refused to allow a production of *Die Liebe der Danae* until after the war, but he yielded to Krauss's persuasion and the première was arranged for the 1944 Salzburg Festival. After the bomb plot against Hitler in July 1944, the festival was cancelled and the *Danae* première became a dress rehearsal on 16 August. Strauss's creative response to the ghastly shambles around him was to compose *Metamorphosen* in spring 1945. This 'study for 23 solo strings' is an elegy for the German musical life of which Strauss had been a leader for half a century. After the surrender of Germany in May 1945 Strauss began the Oboe Concerto, which reflects nothing of the time in which it was written. In October that year he and Pauline went into voluntary exile in Switzerland. Because he had held official office under the Nazi regime Strauss was an automatic candidate for a 'denazification' tribunal. In Zurich and Montreux he was unmolested and heard the first performances of his Oboe Concerto, *Metamorphosen*, the Wind Sonatina no.2 and the Duett-Concertino for clarinet, bassoon and strings. These are all mellow in spirit and wonderfully refined in technique, and are generally known as the works of Strauss's 'Indian summer', a convenient name for his final period, which may be said to have begun with *Capriccio*.

The hand of reconciliation was extended to Strauss by Britain in 1947, when Beecham instigated a Strauss festival in London and invited him to attend. He spent 4–31 October in London, his first visit since 1936 when he had received the gold medal of the Royal Philharmonic Society, and conducted several of his compositions. On his return to Switzerland his health began to fail. In June 1948 he heard that his name had been cleared by the denazification board and he was free to return to Garmisch. First, however, he had a severe operation, and did not go home until May 1949, taking with him four wonderful songs with orchestra, composed during 1948, which were to be his last work and were published posthumously as *Vier letzte Lieder*. He took part in the Munich celebrations of his 85th birthday, but his heart began to fail in August and he died peacefully in Garmisch-Partenkirchen on 8 September 1949. One of his last remarks, utterly characteristic, was made to his daughter-in-law: 'Dying is just as I composed it in *Tod und Verklärung*'.

CHAPTER FIVE

Works: introduction

During his lifetime and afterwards Strauss was the centre of controversy, and even now he remains one of those composers capable of arousing extremes of sympathy or antipathy. There are inevitable variations in the quality of his work, some of it not much more than musical journalism of the kind most court composers turned out as part of their obligations. But the distance between the peak and the base in a graph of Strauss's output is not as wide as it was once believed, and the former critical ordinance that he declined into a state of unimaginative self-repetition after about 1918 is untenable when his achievement is judged in perspective. It is easy to understand how such a view of him became widespread outside Germany, and even inside it. The years from roughly 1890 to 1910 were a brilliant noonday for Strauss, as they were for Elgar in England, when he bestrode the musical world and audiences hung on his every note. He dazzled, he shocked, he amazed. After such a sustained *fortissimo* there was bound to be a diminuendo, and in Strauss's case it was all the steeper because, with the cataclysm of World War I followed by his absorption into the world of the Vienna Staatsoper, he effectively 'disappeared' for a decade. Within that time the face of music changed. Strauss's audacities became the norm; worse, they became outdated clichés in the postwar reaction against Romanticism. A new modernism sprang up. Many strident voices competed for a hearing, Bartók, Stravinsky, Hindemith, Berg, Schoenberg, Satie and Prokofiev among them. It was not that Strauss had nothing left to say within his idiom; people no longer listened to him. His new works rarely penetrated beyond Germany and Austria; opera houses and orchestras contented themselves with the pre-1914 successes. It was as if he were dead.

Strauss saw no reason to change his style to accord with what his friend Romain Rolland called the 'new frenzy' in music. But he was still much concerned with extending his range within the confines of his style and creative personality. This involved a certain amount of experiment, tame by prevailing standards, and not all appreciated at the time. No retrogression was involved, for instance, in the opera *Intermezzo*, which adopted a cinematic stage technique paralleled by mastery of conversational recitative. The *Parergon zur Symphonia domestica* was one pianist's property, but the music had a harmonic toughness and melodic inventiveness which contradicted the self-indulgent confectionery of *Schlagobers*. Each of the operas *Arabella*, *Die schweigsame Frau* and *Daphne* marked a new stage in Strauss's treatment of words and music. Each was in a style appropriate to its content, showing a continual process of refinement which culminated in *Capriccio* and the other masterpieces of his last decade. The clues to Strauss's aesthetic are in two of his pronouncements: 'New ideas must search for new forms' – each of his 15 operas fulfils this precept – and 'Work is a constant and never tiring source of enjoyment to which I have dedicated myself'. This dedication to the 'holy art of music' (*Ariadne auf Naxos*) dominated Strauss's life as conductor and composer. He was a disciplined worker, planning months ahead and keeping to his schedule, goaded by his extraordinary wife. This is not a virtue in itself, nor does it guarantee good results, but because of Strauss's objective, self-denigrating attitude to his work, about which there are many anecdotes, the impression is sometimes given of a casual dilettante. Nothing could be further from the truth. He was completely professional.

Strauss's seemingly relaxed attitude to life, whether he was composing, conducting or playing skat, was a primary source of inspiration. 'Everything I have done casually, with my left hand, has turned out particularly well', he remarked. The elegant *Bourgeois gentilhomme* music was, he said, done with his 'left hand'. Like Elgar, he claimed to think in terms of music at all hours. His sketchbook never left his side; he composed an important theme of *Rosenkavalier* while playing skat. He was unusually and interestingly frank about his methods:

I often write down a motif or a melody, then put it away for a year. When I return to it I find that quite unconsciously something in me – the imagination – has been at work on it. . . . Before I note down even the slightest sketch for an opera I allow the text to permeate my mind for at least six months and take root within me, so that I am wholly familiar with the situations and characters. Only then do I allow musical ideas to enter my head. The preliminary sketches become more elaborate sketches which are written down, worked on, put into shape as a piano score and worked on again, often up to four times. This is the difficult part of the work. I finally write the score in my study, straight through and without effort, working up to twelve hours a day.

Karl Böhm has related how in 1936, when Strauss showed him the libretto of *Daphne*, he noticed that on the margins Strauss had jotted down notes concerning rhythms, tonality and precise indications of the musical form concerning several characters. 'And he had needed', Böhm said, 'scarcely any more time than it took to read through the text.' Strauss was a master of melody: he thought of himself as able to invent only short themes, and many of his wide-spaced operatic melodies are derived from fragmentary phrases. 'I work on melodies for a very long time', he told Max Marschalk, 'what matters is not the beginning of the melody but its continuation, its development into a perfect melodic shape.' Strauss here seems to be supporting the epigram that inspiration means perspiration, yet his ability, which never deserted him, to invent an immediately memorable phrase to depict a character or situation was perhaps his greatest asset as a composer. Examples that come readily to mind are the eruption at the start of *Don Juan* (once described as 'like champagne corks popping'), the *Till Eulenspiegel* motif on the horn, the first bars of *Der Rosenkavalier*, the clarinet arpeggio in C♯ minor that opens *Salome*, the *Don Quixote* theme, the Falcon's cry in *Die Frau ohne Schatten*, and the Child's theme in the *Symphonia domestica*. He was also a thematic kleptomaniac. He made quotation and self-quotation into an art, and many of his motifs are modelled on those of other composers (he would have echoed Vaughan Williams's 'Why should music be original?'). His credo was expressive beauty of sound, vocal and instrumental. He remained true to it, and if there are times when he settled too easily for a cloying sweetness in place of a changing harmonic texture that is an entry on the debit side of an account overwhelmingly in credit.

CHAPTER SIX

Early works

The music that Strauss composed when he was a youth
amply confirms his statement that his father refused to
let him study anything but the classics. The prevalent
influences are Beethoven and Mendelssohn. There is no
doubt about the melodic fertility and facility, but the
disciplines of sonata form were obviously irksome and
unwelcome. The A major String Quartet is a student's
exercise, accomplished but unremarkable. It is significant that the first glimpses of the mature Strauss occur
in the piano pieces *Stimmungsbilder* of 1882–4, where
formal restrictions were abandoned in favour of atmospheric tone-painting. Strauss's struggle for self-
expression while fettered by classical procedures is
eloquently evident in the Cello Sonata, a splendid work
in spite of its formal deficiencies. Joachim congratulated
Strauss on the lyrical opening outburst, with good
reason, for the work, though heavily overlaid with
sequential repetitions, has a vitality that has ensured its
survival. The slightly earlier Violin Concerto is less
ambitious but melodically attractive; it has no cadenza.

Much the best of the works composed in 1881–3,
however, are the one-movement Serenade for 13 wind
instruments (normal double woodwind, four horns and
contrabassoon or optional bass tuba) and the Horn
Concerto no.1. Although Strauss himself later dismissed
the Serenade as 'no more than the respectable work of a

music student', it is easy to hear how its mellifluous grace must have attracted Bülow. Although the work is in sonata form, Strauss avoided his weak point, the development, by substituting an independent central episode in B minor linking exposition and recapitulation. Unity is maintained in this section by pervasive use of a six-note figure derived from the work's second subject. But the outstanding feature is the assured blending and contrasting of instrumental sonorities. The Suite in B♭ for the same combination is in four movements and exposes formal deficiencies, but it is prophetic of Strauss's skill in manipulating short themes and has instrumental felicities, such as the oboe solos, which are pointers to his maturity. It is in the Horn Concerto no.1 that Strauss first fully overcame structural difficulties. Three short movements are played without a break. The solo horn's opening fanfare is metamorphosed into the rondo finale's principal subject, while another horn figure is worked into the texture of each movement. There are other thematic links throughout. An odd feature of the concerto is that while the solo part is written for the valved F horn, the orchestral horn parts are for the valveless E♭ crook. Compared with this delightful and still fresh-sounding concerto, the Symphony in F minor of 1883–4 is a reversion. On a large scale and in cyclic form, it displays Strauss's orchestral mastery in the Scherzo and Andante cantabile; but the contrived recapitulation in the finale of themes from the preceding movements makes a lame conclusion to a work that won high praise at its early performances but was soon to be overtaken by Strauss's progress in new directions. It was in any case overtaken by the contemporaneous Piano Quartet in C minor, also on a large scale but more successful in its formal procedures. The melodic content is heavily indebted to Brahms, but at last Strauss's first-movement development section is logical and sounds spontaneous, and he was in full control of his grandiose design. Therefore the quartet represented a significant advance, consolidated four years later, in 1887, in Strauss's last chamber work and his last 'classical' piece, the Violin Sonata. It is true that this excellent work tends to break the bonds of its format – the violin part has an operatic grandeur and the piano writing suggests orchestration – but it is exciting and rewarding to perform, as Heifetz's lifelong affection for it testifies, and it contains many surprises in the shape of references to Schubert's *Erlkönig* and Beethoven's 'Pathétique' Sonata and in the chromatic modulations in the coda to the finale.

The Violin Sonata was composed later than the two works that marked the true end of the period of apprenticeship, the *Burleske* for piano and orchestra and the symphonic fantasy *Aus Italien*. They are especially significant as signposts to the later Strauss. The *Burleske* began as a Scherzo which Strauss wrote in 1885 for Bülow, who from a sight of the score declared it unplayable while Strauss himself wrote it off as 'nonsense' after a run-through with the Meiningen Orchestra. Not until he showed it in 1890 to Eugen d'Albert, who was enthusiastic, did he give it its present title. It is in extended sonata form, a weakness because it is just a few minutes too long, and there are still Brahmsian echoes, but here for the first time is the witty, sparkling Strauss, playboy of the orchestra, teeming with ideas. Here, too, is the first intrusion of waltz rhythm into a Strauss work. Like Mahler with the march and the ländler,

Strauss was happy to draw elements of popular music into his scheme, and the waltz fulfilled this purpose admirably until it became a mannerism. *Aus Italien* was composed in 1886, and Strauss called it 'the connecting link between the old and new methods'. It is not yet a tone poem, nor, in spite of its four movements, is it a symphony, at least no more than is Berlioz's *Harold en Italie*, the genre to which it belongs. But certainly 'the poetic idea was really the formative element' of this picturesque evocation of the sights, sounds and atmosphere of Strauss's first visit to Italy. The Lisztian first movement, 'Auf der Campagna', is structurally the best and most advanced, but the divided strings of the second movement and the vivid depiction of sunlight on sea in the third by means of 'cascades' for violins and flutes are outstanding pointers to the brilliance of the later tone poems. This third movement is noteworthy in three other respects: as in the wind Serenade an independent episode takes the place of a development; Strauss imitated Liszt's example in following his poetic instinct by juxtaposing the various sections of the movement, eliminating the 'rules' of classical procedure; and in the final bars he first used his favourite device of a harmonic side-slip (i.e. writing as if he was to modulate into a distant key and suddenly sidling into the tonic).

217

CHAPTER SEVEN

Orchestral works

Between 1888 and 1898 Strauss composed the tone poems on which his fame and popularity in the concert hall chiefly rest. The first, *Macbeth*, is experimental and still not mature Strauss. There is no attempt to follow the detailed action of Shakespeare's play, though certain incidents are used. Instead the work is a psychological study. It is in one movement of extended sonata form, in which the lengthy development incorporates two self-contained episodes. *Macbeth* is a striking work, but not as striking as its successors, hence its relative neglect. Its opus number, 23, is later than that of *Don Juan* (op.20), because it was heavily revised and was not performed and published until 1890, a year after *Don Juan*.

Like *Macbeth, Don Juan* is a sonata movement with self-contained episodes, but the difference is remarkable. Under the impact of his love for a divorced woman, Dora Wihan, Strauss turned to Lenau's fragmentary poem which became the basis of a controlled masterpiece into which the fiery ardour of youth was injected with a passion that has never faded. From it dates the appearance of the real Strauss. Its magnificent opening, a theme comprising all the principal elements of the structure, is presented with an orchestral virtuosity that is the strongest evidence of what Strauss had already learnt as a conductor. What he had learnt as a composer is evident in every bar, but nowhere more impressively than in the love scenes, which besides being extremely ardent are distinguished by the broad sweeping cantilena that he was to exploit so tellingly in his operas. Not only had he learnt unprecedented pyrotechnics, at the same time he had mastered the secret of musical continuity.

For *Tod und Verklärung* (1888–9) Strauss provided a detailed synopsis which the music exactly illustrates: an artist, on his deathbed suffering physical agonies, recalls his youth and his unfulfilled idealism; he dies, and his soul achieves transfiguration. The poem by Alexander Ritter printed in the score was written after the music had been composed. The design of this tone poem can still be distantly related to sonata form, divided into slow introduction, symphonic *allegro* and epilogue, the principal themes recurring cyclically. Harmonic dissonances, no longer remarkable, caused a stir in 1890 and contributed to Strauss's growing notoriety as an *enfant terrible*. Although the concluding 'transfiguration' section in C major falls some way short of its sublime target, the harmonic modulations, most imaginatively deployed, are of great beauty if sensitively played.

Five years passed before the appearance of the next tone poem, *Till Eulenspiegels lustige Streiche* (1894–5), described as a rondo (the work's full title in English is 'Till Eulenspiegel's merry pranks, after the old rogue's tale, set for large orchestra in rondo form'). Strauss had contemplated a one-act opera on this subject, but the failure of *Guntram* in 1894 discouraged him. His change of mind was fortunate, though, for in the tone poem the witty and pawky side of Strauss appears fully clad in his new-found virtuosity. *Till* is a masterpiece on every level, as a programmatic description of the rogue's pranks in detail, as a generalized portrait of a scamp, or as an example of musical humour. It is in some respects, too, a self-portrait of Strauss delighting in his affront to the bourgeois philistines of Munich who thought his music so outrageous; he exploited this vein of waspish humour five years later in the opera *Feuersnot*. The thematic transformations in *Till* are ingenious and spontaneous, the scoring deft, picturesque and always apt; and the expansion in the coda-epilogue of the 'once upon a time' introduction is one of those inspired finishing touches, like the end of *Der Rosenkavalier*, in which Strauss specialized. *Till* requires quadruple woodwind, including the D clarinet. The horns were increased to six in *Also sprach Zarathustra* (1895–6), a musical commentary on Nietzsche's poem rather than its programmatic musical equivalent. If Strauss's strong sense of publicity governed his choice of a subject involving the controversial doctrine of the 'superman', it was surely his Bavarian sense of humour that presented the 'Dance of the Superman' as a luscious Viennese waltz. *Zarathustra* marks an advance in Strauss's use of one-movement form, the work being a free fantasia unified by the C–G–C nature motif heard at the outset. The orchestral mastery may be taken for granted, but Strauss's full maturity can be gauged from his confident exploitation of the contrasts between the tonalities of C and B, unresolved even in the final chords. If the use of the 12 notes of the chromatic scale in the slow fugue theme denoting science seems contrived, the still astonishing polytonal effects elsewhere in the work and its beautiful final nocturne have ensured it a secure place in the repertory of great orchestras.

In 1896–7 Strauss composed the most poetic, if not the finest of his orchestral works, *Don Quixote*, 'fantastic variations on a theme of knightly character'. It was an inspiration to use variation form for the Don's adventures; inspired, too, to cast the work as a kind of sinfonia concertante, with solo cello and viola representing (though not exclusively) Don Quixote and Sancho Panza. Strauss's command of musical pictorialism becomes almost arrogantly realistic in *Don Quixote*: windmills, sheep (woodwind and muted brass playing minor 2nds in flutter-tonguing) and the flying horse are as vividly illustrated as by any graphic artist, and it is this aspect of the work which has too often been emphasized at the expense of the extremely subtle psychological portrayal of Don Quixote's unhinged mental state by means of discordant and 'clouded' harmony, resolved only when the Don rides home over a throbbing pedal point.

Strauss regarded his subsequent tone poem, *Ein Heldenleben* (1897–8), as a companion-piece to *Don Quixote*. They were companions in misunderstanding by some commentators. Because it coincided with the superman ethos of Kaiser Wilhelm II's Germany, with the growth of Prussian militarism and the architectural bombast of pre-1914 Berlin, *Ein Heldenleben* was ascribed to Strauss's 'megalomaniac' tendencies – here was a composer writing musical autobiography in terms of superhuman grandiosity, demanding eight horns, five trumpets and quadruple woodwind. But this was the age of the huge orchestra and Strauss, like Mahler, gloried in it. *Ein Heldenleben* is essentially another product of Strauss's Bavarian capacity for self-parody. His hero is no Nietzschean superman but a composer, a

20. The opening of 'Ein Heldenleben', from the autograph
fair copy, begun 2 August 1898

Kapellmeister, whose adversaries are the music critics, who is soothed and cajoled by his capricious wife (represented by the solo violin), whose battle against the critics is halted by his 'works of peace' – his own compositions – and who seeks peace, like Don Quixote, by retiring to the country (with a view of the Alps, as at Garmisch). That is the programme, divided into six sections, and illustrated by music of such inventiveness, humour and homogeneity that it is completely convincing as an abstract composition. Its only rival as musical autobiography on a scale as bizarre as it is undeniably effective is Berlioz's *Symphonie fantastique*.

Autobiography of a more intimate kind is the background of the *Symphonia domestica* (1902–3). Although Strauss removed details of the programme because, like Mahler, he wanted his music to be judged purely as music, the 'domestic symphony' is what its title implies: a picture of Strauss and Pauline at home, quarrelling, bathing the baby, working, loving, dreaming, waking. It is a one-movement symphony in four sections, and valid if regarded as nothing more. Again a vast orchestra is used, including four saxophones, but the scoring is often of chamber music delicacy. The invention is of high quality and the work has generally been underrated. In spite of the complex orchestral apparatus used, there is some simplification of the harmonic style and this is carried further in *Eine Alpensinfonie* (1911–15), in which Strauss described a day in the mountains he could see from his study window in Garmisch. This work is the apotheosis of the Straussian orchestra (over 150 players), a gigantic piece of nature-painting in 22 sections which has a pantheistic exaltation, often expressed in naive diatonic terms, related to that of Mahler's Third Symphony. Of all his great orchestral works it relies least on any knowledge of the programme, and Strauss could claim with justice that he had fully learnt how to merge pictorialism and 'absolute' music into a seamless unified structure. *Eine Alpensinfonie* lacks the pungency of the earlier tone poems and the frenzied orchestral manner of the operas *Salome* and *Elektra*, but its superb contrapuntal texture and the sheer splendour of its sound contradict any suggestion of a creative decline simultaneous with the composition of *Der Rosenkavalier*. The fact that Strauss admitted that he enjoyed composing and said it came easily to him has perhaps engendered a too puritanical resistance by some listeners to such a sumptuous work as *Eine Alpensinfonie*.

Of Strauss's ballet scores, *Josephs-Legende* (1912–14) is the best because it comes nearest to being another tone poem. Strauss confessed difficulty in working up interest in 'good boy Joseph', and his invention became banal when called on to emulate innocence and purity. He could find the music for these qualities in Sophie in *Der Rosenkavalier*, but if they were associated with religion they had no appeal for his muse. *Schlagobers* (1921–2) has little to commend it beyond a glib orchestral expertise decked out garishly. Strauss's arrangements of Couperin were originally made for the ballet stage in 1922 and 1942, but are better known as concert suites; they are among the several works for small orchestra that he wrote in the second half of his life, the first and finest of them being the suite of items salvaged from the incidental music to *Le bourgeois gentilhomme* (1917). The twin influences on Strauss's work were Wagner and Mozart, and his output could almost be divided into Wagner works and Mozart works, with some mixtures (*Der Rosenkavalier*, for example). The *Gentilhomme* music is a Mozart work, though it pays lip-service to Lully. It is no 17th-century pastiche; rather the age of Molière is re-created by a 20th-century artist with incomparable wit and grace. The spirit of Mozart is more explicitly acknowledged in the 'Indian summer' works of 1942–8. The two late wind sonatinas outstay their welcome because Strauss, who admitted he had 'a complicated brain', seems to have been enjoying solving ingenious contrapuntal problems for their own sake. The Oboe Concerto, however, is almost perfect in form and execution, a late swallow of particular charm. But the greatest of the last orchestral works is *Metamorphosen*, a symphonic *adagio* of Mahlerian intensity in which Strauss's inveterate skill in weaving elaborate string textures is no mere 'wrist exercise' but a profound expression of his agony of mind over the destruction of the German culture that had nurtured him. The music's poignancy is heightened by the thematic allusions to Wagner and Beethoven.

CHAPTER EIGHT

Operas

Strauss's operas demonstrate well the development of his style from 1892, when he began to compose *Guntram*, to 1941, when he wrote the magical ending of *Capriccio*. They are his most important contribution to music. A leitmotif of his operatic career was his preoccupation with the clarity of the words, culminating in his using the relative importance of words and music in opera as the theme of his last stage work. His harmonic and tonal procedures deserve close study throughout, especially his predilection for associating certain keys with individuals or situations (e.g. in *Elektra*, B♭ for Agamemnon and E♭ for Chrysothemis). He had worked out his eloquent and agile harmonic idiom in the tone poems, and it stood him in good dramatic stead in the operas. At the start of *Don Juan*, there is an abrupt switch from C major into E major, and at the close of the *Heldenleben* love-music the modulation from G minor to D minor is above a sustained G♭ major chord. The polytonal 'excesses' of *Elektra*, with A major and E♭ minor combined, shocked Debussy because of their 'cold-blooded' audacity. The rich and sustained harmonic texture of the operas from *Rosenkavalier* derives from a technique involving frequent use of multiple suspensions, passing notes and anticipations, and the combination of major and minor elements. The superb arching love-songs are usually in keys with many flats or sharps, D♭ major especially and F♯ major.

Operas

Strauss's first opera, *Guntram* (1892–3), has never held the stage, even in its heavily cut revision of 1940, when Strauss ruefully acknowledged that 'the whole of *Guntram* is a prelude'. He inherited from Wagner the principle of a continuous, seamless texture and a leit-motif system with the orchestra in a dominating role. But only in *Guntram* did he use opera as a propaganda vehicle for a doctrine. It was a work he had to get out of his system, though there is delight to be found in its anticipations of later and greater works, as well as amusement at the plagiarisms from *Tristan* and other Wagner operas. Nor can there be any denial of the mastery of orchestral and vocal techniques; coming after *Don Juan* and *Tod und Verklärung* it can scarcely be called immature. But veneration for Wagner all too often in *Guntram* snuffed out the original Strauss who composed those masterpieces. How well Strauss had learnt his lesson seven years later is apparent in *Feuersnot* (1900–01), which, whatever its faults and difficulties, is a brilliantly effective satire on philistinism executed in Strauss's lighter manner. The seeds of *Der Rosenkavalier* (and its waltzes) can be found throughout *Feuersnot*. The plot was considered shocking in its day and contributed to the impression of Strauss as a scan-dalizer, which received vast reinforcement with the production of *Salome* (1903–5).

Like *Feuersnot*, *Salome* is in one act and, with slight exaggeration, can be called a tone poem with vocal interludes. All the atmospheric power and thematic dex-terity of the tone poems are used in this story of a 16-year-old virgin's perverse obsessions. Naturally the word 'decadence' was much applied to this opera, a tribute to Strauss's skill, perhaps with one eye on the success of *Tosca*, in purveying the flavour of Wilde's *fin-*

227

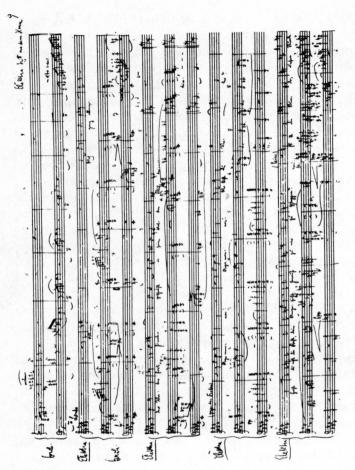

21. Autograph sketch for Electra's monologue

de-siècle Romanticism even in German translation. The instrumental inventiveness of the score is breathtaking, but its sultry beauty, the evocation of the Palestinian night, the vivid delineation of Herod's character and Salome's final 'Liebestod' after her controversial dance are its lasting assets. It is a virtuoso display of the creation of atmospheric colour by instrumental means.

However, its successor, *Elektra* (1906–8), on what is in several respects a similar subject, owes its power to its musical structure, to architecture rather than to painting. In his excitement with his first Hofmannsthal libretto, Strauss matched the gruesome subject of Electra's obsession with revenge with his most advanced music, outdoing anything in *Salome* in dissonance and harmonic waywardness. The harmony derives from a single germinal chord whose flavour pervades the score. The dissonant polytonal episodes, sometimes crossing the border into atonality, are offset by passages of simple diatonicism, which usually arise from Strauss's contrapuntal textures but occasionally are used for a shock effect. A larger orchestra is used than in *Salome*, the wind including eight clarinets of various sorts, a heckelphone and Wagner tubas. Yet this array is controlled and balanced with an extraordinary precision which explains Strauss's *bon mot* that *Salome* and *Elektra* should be conducted 'as if they were fairy music by Mendelssohn'. Controversy continues, not over the dissonances of *Elektra*, which are now familiar enough, but over whether Strauss deliberately retreated into a cosier world from the expressionist anarchy that he had opened up, whether Hofmannsthal failed to provide challenging librettos or whether Strauss himself recognized that he could not repeat *Elektra*. The last seems the most tenable of these suppositions. He remained anxious to discover 'new forms' but they lay along more realistic lines, as in *Intermezzo*.

Not the least of Strauss's achievements in *Salome* and *Elektra* was the magnificent understanding he displayed in exploiting the female voice, something he learnt, presumably, from his wife Pauline, whom he described as the finest interpreter of his lieder. Both operas contain dramatic female roles, Salome and Herodias in the former and Electra, Chrysothemis and Clytemnestra in the latter. Clytemnestra's aria, rewritten three times before Strauss was satisfied, is the first 20th-century musical portrayal of corruption of body and soul, the forerunner of many. In Strauss's next opera, the comedy *Der Rosenkavalier* (1909–10), three very different soprano roles were created that have become touchstones by which operatic reputations are assessed. A legion of great singers since 1911 has revelled in the rewarding challenges of Sophie, Oktavian and the Marschallin, some having sung all three, progressing from the first to the last as their voices have developed. *Der Rosenkavalier* remains the most successful and popular of Strauss's operas. Its superior libretto, its 18th-century Vienna setting, its mingling of romance, farce, wit, sentimentality and tenderness, its human characterizations, its feast of melody dominated by waltz rhythms – all these positive virtues outweigh the occasional longueurs, the excessive, even fussy orchestral detail and the distastefulness of Baron Ochs's humiliation. The work is easily vulgarized, but the key to a successful interpretation can, as always, be found among Strauss's own writings: 'Light, flowing tempi, without compelling the singers to rattle off the text. In a word: Mozart, not

Lehár'. Strauss rarely excelled the vocal opulence of the
Act 3 trio for the three sopranos (Oktavian, a breeches
role, is usually sung by a mezzo, but the score desig-
nates a soprano), and it is typical of Strauss that he
fashioned its soaring melody from a comic phrase sung
earlier in the act by Oktavian disguised as a maid-
servant.

Strauss and Hofmannsthal remained in the 18th century in *Ariadne auf Naxos* (1911–12, 1916). The first
version, with the Molière play performed before the
Ariadne opera, is rarely heard; the more familiar 1916
version with the felicitous prologue contains some of
Strauss's best music and also the character of the
Composer (sung by an Oktavian soprano). Although it
is comparatively short, this role has attracted leading
singers from Lotte Lehmann onwards because it is so
satisfying to perform. Hofmannsthal's happy conception
of a mingling of *commedia dell'arte* characters with the
tragedians of *Ariadne* appealed to Strauss, after initial
hesitations, and gave him free rein for a combination of
the Mozart and Wagner sides of his personality (or the
Eulenspiegel and *Heldenleben* sides). The prologue, as
full of action as the levée in *Rosenkavalier* and juxtaposing sentiment and force in a manner highly agreeable to
Strauss, rarely fails in performance; the *Ariadne* opera
itself has come into its own when intelligent production
combined with well-rehearsed singing and acting has
revealed the excellence of the conjuring-trick Strauss
performed in his juxtaposition of frivolity and serious-
ness, epitomized by Zerbinetta's sparkling coloratura
aria and Ariadne's solemn and entrancing 'Es gibt
ein Reich'. The role of Bacchus requires a Helden-
tenor who must look as well as sing like a god, a

22. Hugo von Hofmannsthal and Strauss, c1915

combination rarely found on the opera stage. Deficiencies in either respect can damage the effect of the final duet, but in a great performance (several have been recorded) doubts about the quality of the music are stilled. The use of a small orchestra (37 instruments) is one of the work's happiest features, revealing Strauss as a master of small-scale sonorities yet never sounding thin or etiolated. The second 'Grecian' opera by Strauss and Hofmannsthal, *Die ägyptische Helena* (1923–7), was marred by a comparable but more damaging divergence of aim between composer and librettist. As early as 1916, in the middle of the war, Strauss had told Hofmannsthal that 'm˙ tragic vein is more or less exhausted . . . I feel downrighı called upon to become the Offenbach of the 20th century. . . . Sentimentality and parody are the sensations to which my talent responds most forcefully and productively'. He added, engagingly: 'After all, I'm the only composer nowadays with some real humour and a sense of fun and a marked gift for parody'. So when the collaborators decided to write an opera about Helen of Troy, the project started out lightheartedly, with Strauss composing the first act just as fluently as he had *Rosenkavalier*. But in Act 2 Hofmannsthal began to philosophize and to complicate, with the inevitable result that the work can seem broken-backed and has flaws. Strauss's music, composed for a legendary group of Vienna singers, is in his equivalent of a bel canto style, and the refined yet luscious orchestration is extraordinarily persuasive.

Between *Ariadne* and *Helena*, Strauss had composed two of his most important operas, *Die Frau ohne Schatten* (1914–18) with Hofmannsthal, and the 'bourgeois comedy' *Intermezzo* (1918–23) to his own libretto. The interruptions of war prolonged the genesis of *Die Frau* to a point where Strauss called it his 'child of sorrow'. The libretto is Hofmannsthal's most symbolic and intellectual creation, a mixture of fairy tale, magic and Freudian psychology. Strauss, as devil's advocate on behalf of the audience, steered his librettist away from some of his more oblique ideas, and once confessed himself unable to find 'red corpuscles' in the characters. Yet he recognized that it was Hofmannsthal's finest achievement and some think that it is also Strauss's. There are good reasons to hold this opinion. No work of his, not even *Don Quixote*, is more memorably scored, a large orchestra being used with a Mahlerian virtuosity in contrasts of opulent exoticism and chamber music intimacy (the influence of *Das Lied von der Erde* is detectable). The roles of the Empress, the Dyer's Wife and the evil Nurse again challenged Strauss to produce outstanding music for female voices, but he was unexpectedly successful with the tenor role of the Emperor and even more with the delightfully human and warm bass-baritone part of Barak the Dyer (one of several roles from which it is possible to deduce a musical self-portrait). As a contrast from the high-flown Romanticism of this masterpiece Strauss enjoyed the relaxation of creating *Intermezzo*, which marks a radical change in his style (at a time when he was supposed to be extinct). The musical aspects of *Intermezzo* were obscured in its first years by the publicity surrounding its plot, a blatant slice of the Strausses' real life when Pauline falsely accused Richard of adultery. Domestic quarrels, rows with the cook, skiing, a lullaby for the child, a skat game, reconciliation, all are vividly re-created in a taut structure comprising a succession of short scenes linked by orchestral interludes (the parallel with *Wozzeck* is apparent and coincidental). The novel use of *secco* recitative was dictated by Strauss's preoccupation with the audibility of the text. In *Intermezzo*, he warned conductors, 'all passages of pure dialogue – in so far as they do not change for short periods of time into lyrical outpourings – in other words, all passages resembling *recitativo secco*, should be presented *mezza voce* throughout'. In its insistence on a naturalistic style of operatic performance, *Intermezzo*, this 'harmless comedy' as Strauss called it, was ahead of its day. It is a more serious piece than it seems – as the sensitive Hofmannsthal realized when he saw it – and it occupies a crucial place in Strauss's development of conversational operatic dialogue between the *Ariadne* prologue and the near-perfection he attained in *Capriccio*; and no doubt he would have been the first to acknowledge his debt to the Sachs–Eva scene in *Die Meistersinger*. Yet the piece was not taken seriously until 50 years after it was written. Compared with the realism of *Wozzeck*, the 'opera domestica' of Strauss was altogether too bourgeois a comedy for its day.

The last of the Hofmannsthal operas, *Arabella* (1930–32), is a strange, flawed work. This attempt at a 'second *Rosenkavalier*', an opera it in no way resembles except in its Vienna setting of a century later, is handicapped because the libretto of the second and third acts was left unrevised as a memorial tribute to Hofmannsthal, whereas the number of alterations to Act 1 made at Strauss's behest is a pointer to the need for later improvements. (Many of the best dramatic effects in their operas were Strauss's ideas.) As if to disguise

235

the obvious weaknesses of plot, Strauss wrote an especially euphonious score, with orchestration of particular delicacy and variety of colour. Paradoxically he transformed his librettist's effort to provide him with the operetta he craved into a full-scale operatic portrait of the heroine. Her love-duet with Mandryka, her Croatian suitor, is among Strauss's most deeply felt creations in this genre. The lighter aspects of the opera benefit from the dialogue style of *Intermezzo*, but the rather obvious emulation of *Die Fledermaus* in Act 2 can wear thin in all but a supreme production. Yet the undeniable charm of the work, its mixture of sympathetic and bizarre characters, appeals to the public, is theatrically effective and makes *Arabella* a pleasing entertainment. It pleases more than its successor *Die schweigsame Frau* (1933–4). This was Strauss's first opera with Zweig, and he said that its composition came more easily than any of his previous operas. It was also the libretto that he altered least. It is his only 'Italian opera', a *buffa* work paying homage to Rossini and to the Verdi of *Falstaff*, the character of Sir Morosus being modelled on that of the Shakespeare–Verdi knight. Ben Jonson's London is updated to 1780 in order that an opera company can be introduced, providing Strauss with the opportunity he so enjoyed for working quotations into the score. There is a music-lesson scene (based on a Monteverdi theme) and several elaborate ensembles. The vocal writing for the soprano Aminta and the high tenor Henry is of extreme difficulty but very effective. The clever score is full of 'gems', sparkling and genuine, but the total effect fails fully to embody the composer's gleeful enjoyment, principally because the melodic invention is not of prime Straussian cut.

For his next two operas, with librettos by Gregor supervised by Zweig, Strauss reverted to the one-act form of his pre-war successes. *Friedenstag* (1935–6), set in a beleaguered fortress during the Thirty Years War, is the most austere of his operas and the finest of those that are little known and underrated. Never loath to acknowledge his dependence on models – *Meistersinger*, *Figaro*, *Zauberflöte* and *Il barbiere di Siviglia* among them – Strauss here ventured into the contrasts of darkness and light, war and freedom, that characterize *Fidelio*. It was a brave work to write in 1936 in Nazi Germany but, politics apart, its harmonic strength, ambitious use of the chorus, reliance on mainly male voices and Mahlerian juxtaposition of the popular and the esoteric overcome defects in Gregor's libretto and compel ungrudging admiration for a septuagenarian's elasticity of approach to operatic form. Its companion-piece, *Daphne* (1936–7), is in total contrast filled with pastoral lyricism, autumnal tints and an impression of warm sunlight captured, it seems, from the Italian surroundings in which most of it was composed. It was unusual for Strauss to write for two tenors, and the roles of Apollo and Leukippos are skilfully and sympathetically drawn; but the singer of Daphne is blessed with one of Strauss's most appealing female roles, demanding a sustained, light, lyrical line combined with occasional dramatic weight. Monotony is avoided by the Dionysian dances and by a command of orchestration by now so magisterial that it was mistaken for habit. Yet no mere habit could have produced the F♯ major iridescence of the final transformation scene nor of the 'shower of gold' episode in the next and last Gregor opera *Die Liebe der Danae* (1938–40). This was Strauss's last excursion into the opulence of *Die Frau ohne Schatten*, a 'cheerful mythology' in three acts which began as a lighthearted idea by Hofmannsthal in 1920 and was expanded by Gregor, at Strauss's insistence, becoming weightier in the process. Yet for all the sneers at Gregor because of the undoubted obscurity of some points in the libretto, the fact remains that Strauss set it and considered that some of the music was as fine as he had ever written, as indeed it is. The score requires a producer and conductor who can bring its extraordinary luminosity to full life in a manner to match the old composer's touchingly noble and festive celebration of all that the Greek classical tradition had meant to him. The solo parts for Jupiter (bass-baritone) and Danae (soprano) are melodically and harmonically as rich as any Strauss created, while the quartet for Jupiter's four ex-mistresses and the structural strength of other ensembles deserve the adjective 'Olympian'. His reluctance to allow this, the last of his big-scale works, to be performed because of his experiences with *Die Frau ohne Schatten*, is an eloquent admission that he knew how much it depended for full success on an imaginative and extravagant production in tune with its multifaceted mood.

The last opera was *Capriccio* (1940–41), to a libretto by Clemens Krauss which owed much to assistance from Hans Swarowsky, Strauss himself, Zweig and even Hofmannsthal. It is described as a 'conversation piece', the term Strauss had applied to *Intermezzo* in his preface to that opera in 1924. *Capriccio* might so easily have been an old man's indulgence, an opera with, as its basic idea, an unresolved discussion on Strauss's favourite topic of the relative importance in opera of words

23. *Final page of autograph score of 'Capriccio', dated 3 August 1941*

and music – a long way, this intellectual subject, from the solar-plexus realities of *Elektra*. Yet by setting the scene in a pre-Revolution French château, the house of a beautiful young widowed Countess and her brother, by

239

symbolizing the problem through the rivalry for the Countess's love of a poet and a musician, by introducing the character of a theatre director with elements of Reinhardt in his make-up and by the crowning inspiration (by the Count) that poet and musician should compose an opera about the events they are enacting, Krauss provided Strauss, through this marvellous one-act libretto, with what he had for so long craved, 'a second *Rosenkavalier*, without the longueurs'. Without the coarse humour, too, and without the waltzes, yet with the chances for parody (there is even another Italian aria), with the equivalent of another levée scene, and, best of all, with a central female character who embodies features of the Marschallin, Arabella, Danae and Daphne and is more admirable than any of them. Strauss took the bait. None of his opera scores is more refined, more translucent, more elegant, more varied and none ends so magically, with a long soliloquy for the Countess in which Strauss's melodic vein and consummate stagecraft show no diminution in their capacity to enslave an audience. If *Capriccio* perhaps lacks the dramatic weight of earlier works, it excels them in the sheer art of economical composition. In this respect, and in its place at the end of a long line of theatrical explorations, it can compare with Verdi's *Falstaff*.

CHAPTER NINE

Choral music and lieder

If only a small number of Strauss's 200 or so songs is well known, the plight of his choral works is worse. Yet the vocal equivalent of the orchestral wizardry of his tone poems is to be found in the 16-part complexities of the *Deutsche Motette* (1913, Rückert) and of the *Zwei Gesänge* op.34 (1897), *Der Abend* (Schiller) and *Hymne* (Rückert). Of particular interest, from the end of his career (1943), is *An den Baum Daphne*, a difficult but magnificent setting of Gregor's original choral finale to the opera *Daphne*, which Strauss discarded in favour of the orchestral transformation scene. Incidentally, the principal motif of *Daphne* is derived from the first theme of the 1935 setting for double chorus of Rückert's *Die Göttin im Putzzimmer*.

The bulk of Strauss's output of lieder was composed between 1885 and 1906. Throughout this time his wife, Pauline de Ahna, was professionally active and many of the songs were written for her to sing, with Strauss as her accompanist. Others were written with particular favourite performers in mind, for example Elisabeth Schumann and Paul Knüpfer. Strauss's outstanding contribution to the development of the lied was his continuation of the Berlioz–Wagner–Mahler style of song with orchestra. He orchestrated many of his songs and sanctioned orchestrations by others, and in some cases the accompaniment was originally composed for orchestra.

He was as unselective as Schubert in his choice of texts, provided they generated the impulse for music in him. He described his methods candidly:

Musical ideas have prepared themselves in me – God knows why – and when, as it were, the barrel is full, a song appears in the twinkling of an eye as soon as I come across a poem more or less corresponding to the subject of the imaginary song. . . . If I find no poem corresponding to the subject which exists in my sub-conscious mind, then the creative urge has to be re-channelled to the setting of some other poem which I think lends itself to music. It goes slowly, though. . . . I resort to artifice.

The melodic lyricism of Strauss's style, so evident in his operas, is no less marked in his songs and burst forth in 1885, when he was 21, in his Gilm settings op.10. These include the ornate setting of *Allerseelen* and the masterly *Die Nacht*, which he hardly ever excelled for creation of atmosphere by tonal ambiguity. It is possibly an even finer song than the justly celebrated *Morgen!* op.27 no.4 of 1894, an example of that simplicity which the 'complicated brain' of Strauss could always produce when he was in the mood. *Morgen!* was one of the first of his songs to be orchestrated (1897), and he preserved its mood of rapture by delicate scoring for three horns, harp and strings, including solo violin. *Cäcilie*, a passionate love-song from the same set, was also orchestrated in 1897, when Strauss transposed the E major piano accompaniment to a heroic E♭. Curiously, the first song of this group, *Ruhe, meine Seele*, was not orchestrated until 1948, when Strauss conjured a transparent sound from huge forces but, perhaps deliberately, changed the mood of the song from rapt stillness to pessimistic gloom. As one would expect from his operas, Strauss was at ease with the 'character' song, especially when a bantering or mocking tone was required. Good examples of this are *Wozu noch, Mädchen* op.19 no.1, a Schack setting, *All' mein*

Gedanken op.21 no.1, to words by Dahn, and *Muttertändelei* op.43 no.2 (Bürger). The superb cantilena that is a Straussian hallmark is to be heard in *Morgen!*, in *Freundliche Vision* op.48 no.1 (Bierbaum), a revolutionary song in its day (1900) because the voice part did not appear to have any connection with the piano accompaniment, and in *Traum durch die Dämmerung* op.29 no.1 (Bierbaum), a song in which Strauss uncharacteristically altered the key signature from F♯ major to B♭ when the poem switches from a description of twilight to a description of love. Among the greatest of his songs are the six Brentano settings of op.68 (1918), wide-ranging in mood, challenging in technical difficulty and completely apposite in word-setting. From 1918, too, comes a curiosity, the *Krämerspiegel* op.66, 12 settings of satirical verses by Alfred Kerr which make scurrilous references by means of puns to various publishers. The hiatus between 1906 and 1918 in composition of lieder was only partly caused by opera composition; it is also attributable to a lengthy wrangle over copyright of songs which came to a head when Bote & Bock threatened Strauss with an action for breach of contract. Strauss took his revenge in these songs and offered them to Bote & Bock (who refused them). They are far from negligible, but the cycle's importance is in the melody of the piano introduction to no.8, which Strauss 'borrowed' over 20 years later as the most romantic and evocative melody in *Capriccio*. If that was a case of recovery of buried treasure, there are other treasures to be found among Strauss's songs, notably in the 'radical' settings of Dehmel, such as *Der Arbeitsmann* op.39 no.3, and the Rückert songs of op.46.

Fittingly, and with all the flair for bringing down the

24. Richard Strauss, 1947

curtain at the right moment which distinguished his sense of theatre, learnt all those years ago at Meiningen, Strauss ended his composing career with the *Vier letzte Lieder* with orchestra which have become, deservedly, loved by the public and admired by the connoisseur. These 1948 settings of three poems by Hesse and one by Eichendorff are not only extremely beautiful; they continue the vein of introspection that had been opened in *Metamorphosen*, or even earlier in the 1935 Rückert song for bass voice, *Im Sonnenschein*. Gloriously written for the soprano voice, the cantilena as cunningly spread as ever, the harmonies as ensnaring, the melody as richly suggestive of the halcyon days of German song, the orchestration as gorgeous as in *Arabella* yet as discreet as in *Capriccio*, these songs have a solemn profundity that makes them an appositely contrived ending to the career of a composer who compensated for what he lacked in spirituality by his astonishing insight into the human heart. The poems ask 'Is this perhaps death?' (and Strauss quoted from *Tod und Verklärung*) and speak of summer closing its weary eyes. No better end could be imagined to the 50 years of musical autobiography by this entertainer of the public who never made the mistake of either exaggerating or underestimating his flawed but generous capabilities. On his last visit to London in 1947 he remarked: 'I may not be a first-rate composer, but I *am* a first-class second-rate composer', perhaps consciously echoing Verdi's remark about himself – that he may not have been a great composer but he was a very experienced one. Strauss was making the same modest claim, but he can no more be denied the greatness than can Verdi.

WORKS

Works without op.no are given the no. assigned them in E. H. Mueller von Asow: *Richard Strauss: thematisches Verzeichnis* (Vienna, 1959–74), iii [AV]

Numbers in the right-hand column denote references in the text.

OPERAS
195–201, 226–40

op.		
25	Guntram (3, Strauss), 1892–3: Weimar, Court Theatre, 10 May 1894; rev. 1934–9, Weimar, Deutsches Nationaltheater, 29 Oct 1940	98, 192, 193, 195, 200, 219, 226, 227
50	Feuersnot (1, E. von Wolzogen), 1900–01; Dresden, Court Opera, 21 Nov 1901	103, 195, 220, 227
54	Salome (1, Wilde, trans. H. Lachmann), 1903–5; Dresden, Court Opera, 9 Dec 1905	104, 195, 196, 198, 213, 224, 227, 229, 230
58	Elektra (1, Hofmannsthal), 1906–8; Dresden, Court Opera, 25 Jan 1909	197, 198, 224, 226, 228, 229, 230, 239
59	Der Rosenkavalier (3, Hofmannsthal), 1909–10; Dresden, Court Opera, 26 Jan 1911	196, 197, 198, 200, 203, 212, 213, 220, 224, 225, 226, 227, 230–31, 233, 235, 240
60	Ariadne auf Naxos (1, Hofmannsthal), 1911–12; Stuttgart, Court Theatre, 25 Oct 1912; 2nd version 1916, Vienna, Court Opera, 4 Oct 1916	198, 199, 211, 231, 233, 235
65	Die Frau ohne Schatten (3, Hofmannsthal), 1914–18; Vienna, Staatsoper, 10 Oct 1919	200, 202, 213, 233, 234, 238
72	Intermezzo (2, Strauss), 1918–23; Dresden, Staatsoper, 4 Nov 1924	201, 203, 211, 230, 234–5, 23c, 238
75	Die ägyptische Helena (2, Hofmannsthal), 1923–7: Dresden, Staatsoper, 6 June 1928; Act 2, rev. (L. Wallerstein), 1932–3; Salzburg, Festspielhaus, 14 Aug 1933	203, 233
79	Arabella (3, Hofmannsthal), 1929–32; Dresden, Staatsoper, 1 July 1933	203, 211, 235–6, 245
80	Die schweigsame Frau (3, Zweig, after Jonson), 1933–4; Dresden, Staatsoper, 24 June 1935	204, 205, 206, 211, 236
81	Friedenstag (1, Gregor), 1935–6; Munich, Staatsoper, 24 July 1938	206, 237
82	Daphne (1, Gregor), 1936–7; Dresden, Staatsoper, 15 Oct 1938	206, 211, 212, 237, 241
83	Die Liebe der Danae (3, Gregor, after Hofmannsthal), 1938–40; Salzburg, Festspielhaus, 16 Aug 1944 (dress rehearsal for cancelled première); Salzburg, Festspielhaus, 14 Aug 1952	206, 207, 208, 237–8
85	Capriccio (1, Krauss), 1940–41; Munich, Staatsoper, 28 Oct 1942	207, 208, 211, 226, 235, 238–40, 243, 245

OTHER STAGE WORKS

—	Romeo und Julia (incidental music, Shakespeare), AV86; Munich, 23 Oct 1887	
60	Der Bürger als Edelmann: Le bourgeois gentilhomme) [incl. frag. from Lully], 1912; Stuttgart, Hoftheater, 25 Oct 1912; rev. 1917; Berlin, Deutsches Theater, 9 April 1918	198, 200, 212, 224, 225
63	Josephs-Legende (ballet, 1, H. Kessler, Hofmannsthal), 1912–14; Paris, Opéra, 14 May 1914	200, 224
70	Schlagobers (ballet, 2, Strauss), 1921–2; Vienna, Staatsoper, 9 May 1924	203, 211, 224
—	Verklungene Feste (ballet, P. and P. Mlakar), AV128, 1940 [Tanzsuite, AV107, orch, 1923, with 6 new nos. later incl. in Divertimento, op.86], 1941; Munich, Bayerische Staatsoper, 5 April 1941	218–25

ORCHESTRAL

—	Overture for Singspiel Hochlands Treue, AV15, 1872–3	
—	Concert Overture, b, AV30, 1876	
1	Festmarsch, Eb, 1876	186
—	Serenade, G, AV32, 1877	
—	Overture, E, AV51, 1878	
—	Romanze, Eb, AV61, cl, orch, 1879	
—	Overture, a, AV62, 1879	
—	Symphony, d, AV69, 1880	186
4	Suite, Bb, 13 wind, 1884	187, 215
7	Serenade, Eb, 13 wind, 1881	187, 214–15, 217
8	Violin Concerto, d, 1880–82	186, 214
11	Horn Concerto no.1, Eb, 1882–3	187, 188, 214, 215
—	Romanze, F, AV75, vc, orch, 1883	
—	Lied ohne Worte, Eb, AV79, 1883	
—	Concert Overture, c, AV80, 1883	
12	Symphony, f, 1883–4	

29 Drei Lieder (Bierbaum), 1895: Traum durch die Dämmerung, 243
Schlagende Herzen, Nachtgang
— Wir beide wollen springen (Bierbaum), AV90, 1896
31 Drei Lieder: Blauer Sommer (Busse), 1896; Wenn (Busse), 1895;
Weisser Jasmin (Busse), 1895; Stiller Gang (Dehmel) [added
no.], with va, 1895
32 Fünf Lieder, 1896: Ich trage meine Minne (Henckell), Sehnsucht
(Liliencron), Liebeshymnus (Henckell), O süsser Mai (Henckell),
Himmelsboten (Des Knaben Wunderhorn)
33 Vier Gesänge, 1v, orch: Verführung (Mackay), 1896; Gesang der
Apollopriesterin (E. von und zu Bodman), 1896; Hymnus, 1896;
Pilgers Morgenlied (Goethe), 1897
36 Vier Lieder: Das Rosenband (Klopstock), 1897: Für fünfzehn
Pfennige (Des Knaben Wunderhorn), 1897; Hat gesagt – bleibt's
nicht dabei (Des Knaben Wunderhorn), 1898; Anbetung
(Rückert), 1898; no.1 orchd 1897
37 Sechs Lieder: Glückes genug (Liliencron), 1898; Ich liebe dich 243
(Liliencron), 1898; Meinem Kinde (Falke), 1897; Mein Auge
(Dehmel), 1898; Herr Lenz (Bodman), 1896; Hochzeitlich Lied
(A. Lindner), 1898; no.2 orchd 1943; no.3 orchd 1897; no.4
orchd 1933
39 Fünf Lieder, 1898: Leises Lied (Dehmel), Junghexenlied 243
(Bierbaum), Der Arbeitsmann (Dehmel), Befreit (Dehmel), Lied
an meinen Sohn (Dehmel); no.4 orchd 1933
41 Fünf Lieder, 1899: Wiegenlied (Dehmel), In der Campagna
(Mackay), Am Ufer (Dehmel), Bruder Liederlich (Liliencron),
Leise Lieder (Morgenstern); no.1 orchd ?1916
43 Drei Lieder, 1899: An Sie (Klopstock), Muttertändelei (G. A. Bür- 243
ger), Die Ulme zu Hirsau (Uhland); no.2 orchd 1900
44 Zwei grössere Gesänge, A/B, orch, 1899: Notturno (Dehmel),
Nächtlicher Gang (Rückert)
— Weihnachtsgefühl (Greif), AV94, 1899
46 Fünf Lieder (Rückert): Ein Obdach gegen Sturm und Regen, 1900; 243
Gestern war ich Atlas, 1899; Die sieben Siegel, 1899;
Morgenrot, 1900; Ich sehe wie in einem Spiegel, 1900
47 Fünf Lieder (Uhland), 1900: Auf ein Kind, Des Dichters
Abendgang, Rückleben, Einkehr, Von den sieben Zechbrüdern;
no.2 orchd 1918
48 Fünf Lieder, 1900: Freundliche Vision (Bierbaum), Ich schwebe 243
(Henckell), Kling! (Henckell), Winterweihe (Henckell),
Winterliebe (Henckell); nos.1, 4 and 5 orchd 1918
49 Acht Lieder: Waldseligkeit (Dehmel), 1901; In goldener Fülle (P.

Remer), 1901; Wiegenliedchen (Dehmel), 1901; Das Lied des
Steinklopfers (Henckell), 1901; Sie wissen's nicht (O. Panizza),
1901; Junggesellenschwur (Des Knaben Wunderhorn), 1900;
Wer lieben will (C. Mündel: Elsässische Volkslieder), 1901; Ach,
was Kummer, Qual und Schmerzen (Mündel: Elsässische
Volkslieder), 1901; no.1 orchd 1918
51 Zwei Gesänge, B, orch: Das Thal (Uhland), 1902; Der Einsame
(Heine), 1906; no.2 also in pf version
56 Sechs Lieder: Gefunden (Goethe), 1903; Blindenklage (Henckell),
1903-6; Im Spätboot (Meyer), 1903-6; Mit deinen blauen
Augen (Heine), 1903-6; Frühlingsfeier (Heine), 1903-6; Die
heiligen drei Könige aus Morgenland (Heine), 1903-6; no.5
orchd 1933, no.6 orchd 1906
— Der Graf von Rom (textless), AV102, 2 versions, 1906
66 Krämerspiegel (A. Kerr), 1918: Es war einmal ein Bock; Einst kam 243
der Bock als Bote; Es liebe einst ein Hase; Drei Masken sah ich
am Himmel stehn; Hast du ein Tongedicht vollbracht; O lieber
Künstler sei ermahnt; Unser Feind ist, grosser Gott; Von
Händlern wird die Kunst bedroht; Es war mal eine Wanze; Die
Künstler sind die Schöpfer; Die Händler und die Macher; O
Schöpferschwarm, o Händlerkreis
67 Sechs Lieder, 1918: I Lieder der Ophelia (Shakespeare, trans. K.
Simrock); Wie erkenn ich mein Treulieb vor andern nun?; Guten
Morgen, 's ist Sankt Valentinstag; Sie trugen ihn auf der Bahre
bloss; II Aus den Büchern des Unmuts der Rendsch Nameh
(Goethe): Wer wird von der Welt verlangen; Hab' ich euch denn
je geraten; Wanderers Gemütsruhe
68 Sechs Lieder (Brentano), 1918: An die Nacht; Ich wollt' ein 243
Sträusslein binden; Säusle, liebe Myrthe; Als mir dein Lied er-
klang; Amor; Lied der Frauen; nos.1-5 orchd 1940, no.6 orchd
1933
69 Fünf kleine Lieder, 1918: Der Stern (A. von Arnim), Der Pokal
(Arnim), Einerlei (Arnim), Waldesfahrt (Heine), Schlechtes
Wetter (Heine)
— Sinnspruch (Goethe), AV105, 1919
71 Drei Hymnen von Friedrich Hölderlin, S/T, orch, 1921: Hymne an
die Liebe, Rückkehr in der Heimat, Die Liebe
— Durch allen Schall und Klang (Goethe), AV111, 1925
77 Gesänge des Orients (trans. Bethge), 1928: Ihre Augen (Hafiz),
Schwung (Hafiz), Liebesgeschenke (Die chinesische Flöte), Die
Allmächtige (Hafiz), Huldigung (Hafiz)
— Wie etwas sei leicht (Goethe), AV116, 1930

9 Stimmungsbilder, 1882–4: Auf stillen Waldespfad, An einsamer 187, 214
    Quelle, Intermezzo, Träumerei, Heidebild
—   Improvisationen und Fuge über ein Originalthema, AV81, 1884
    [only fugue survives]
—   Parade-Marsch des Regiments Königs-Jäger zu Pferde no.1, AV97,
    1905
—   Parade-Marsch Cavallerie no.2, AV98, 1905
—   De brandenburgsche Mars, AV99, 1905–6
—   Königsmarsch, AV100, 1906

ARRANGEMENTS, ETC
C. W. Gluck: Iphigénie en Tauride, 1899; Weimar, Hoftheater, 9 June
    1900
L. van Beethoven: Die Ruinen von Athen (text rev. Hofmannsthal),
    1924, Vienna, Staatsoper, 20 Sept 1924
W. A. Mozart: Idomeneo (text rev. L. Wallerstein), 1930; Vienna,
    Staatsoper, 16 April 1931
Cadenza for Mozart's Pf Conc. k491, 1885   189
Principal publishers: Boosey & Hawkes, Fürstner, Universal

WRITINGS
W. Schuh, ed.: Richard Strauss: Betrachtungen und Erinnerungen
    (Zurich, 1949, enlarged 2/1957, 3/1981; Eng. trans., 1953/R1974)
    [collected essays]
E. Krause, ed.: Richard Strauss Dokumente: Aufsätze, Aufzeichnungen,
    Vorworte, Briefe, Reden (Leipzig, 1980)
    * — ed. in Schuh

    (Asow nos. given in parentheses)

Aus Italien [analysis] (307), 1889; *Tannhäuser-Nachklänge (308),
1892; Tagebuch der Griechenland- und Ägyptenreise (309), 1892;
Über mein Schaffen (310), 1893; Rundschreiben über die Parsifal-
Schutzfrage (311), 1894; Handschreiben zur Reform des
Urheberrechtsgesetzes von 1870 (312), 1898; Autobiographische
Skizze (313), 1898; *Einleitung zu Die Musik: Sammlung illustrier-
ter Einzeldarstellungen (314), 1903; *ed. and enlarged:
Instrumentationslehre von Hector Berlioz (316–17), 1904–5; Zum
Tonkünstlerfeste: Begrüssung anlässlich des Tonkünstlerfestes des
Allgemeinen Deutschen Musikvereins (318), 1905; Bemerkungen
über amerikanische Musikpflege (319), 1907; *Gibt es für die Musik
eine Fortschrittspartei? (320), 1907; Salomes Tanz der sieben

Schleier (320A), ?1908; Zum Salome-Verbot in Amerika (320B),
1908; Rechtfertigung der Aufführung der Symphonia domestica im
Warenhaus Wannemaker (320C), 1908; Elektra: Interview für den
Berliner Lokalanzeiger (320D), 1908
*Geleitwort zu Leopold Schmidt Aus dem Musikleben der Gegenwart
(321), 1908; Die hohen Bach-Trompeten (322), 1909; *Persönliche
Erinnerungen an Hans von Bülow (324), 1909; Dementi zu falschen
Pressenotizen über die neue Oper Ochs von Lerchenau (Der
Rosenkavalier) (327), 1909 or 1910; *Gustav Mahler (328), 1910;
Die Grenzen des Komponierbaren (329), 1910; Der Rosenkavalier
(330), 1910; *Mozarts Così fan tutte (331), 1910; Erwiderung auf
Angriffe gegen den 'Programm-Musiker' (331A), 1911; Antwort auf
die Rundfrage 'Worin erblicken Sie die entscheidende Bedeutung
Franz Liszts für die Entwicklung des deutschen Musiklebens?'
(331B), 1911; *Zur Frage des Parsifal-Schutzes: Antwort auf eine
Rundfrage (332), 1912; *Offener Brief an einen Oberbürgermeister
(333), 1913; *Städtebund-Theater: eine Anregung (334), 1914; Eine
Kundgebung Richard Strauss' (335), 1919; Begrüssungsansprache
des neuen Wiener Operndirektors Richard Strauss (336), 1919; *The
Composer Speaks (in D. Ewen: The Book of Modern Composers,
1945, p.54 (336A), 1921; *Novitäten und Stars: Spielplaner-
wägungen eines modernen Operndirektors (338), 1922; Einleitung
zu Schlagobers (338), 1924; *Vorwort zu Intermezzo (2 versions)
(339–40), 1924; *Über Johann Strauss (341), 1925; *Gedächtnis-
rede auf Friedrich Rösch (342), 1925
*Die 10 goldenen Lebensregeln des hochfürstlichen bayerischen
Kammerkapellarius Hans Knappertsbuch, Monachia (343), 1927;
Der Dresdner Staatsoper zum Jubiläum (344), 1928; *Interview über
Die ägyptische Helena (345), 1928; *Die Münchener Oper (346),
1928; Über Schubert: ein Entwurf (347), ?1928; *Über
Komponieren und Dirigieren (348), 1929; Die schöpferische Kraft
des Komponisten (349), 1929; *Vorwort zu Hans Diestel: Ein
Orchestermusiker über das Dirigieren (350), 1931; Die ägyptische
Helena (351), 1931; Über den musikalischen Schaffensprozess (352),
1931; Gedenkworte für Alfred Roller (353), 1933; Über Richard
Wagner (354), 1933; An die Schriftleitung Musik im Zeitbewusstsein
(355), 1933; *Zeitgemässe Glossen für Erziehung zur Musik, für
einen befreundeten Pädagogen (356), 1933; Ansprache des Präsi-
denten der Reichsmusikkammer Dr. Richard Strauss anlässlich
der Eröffnung der ersten Arbeitstagung der Reichsmusikkammer
(357), 1934; *Appell zum 'Schutz der idealen Interessen am

Kunstwerk' (358), 1933 or 1934. Ansprache bei der öffentlichen Musikversammlung am 17. Februar in der Berliner Philharmonie (359), 1934; Ansprache am ersten Komponistentag in Berlin (359A), 1934; Zur Urheberrechtsreform (359B), 1934; Musik und Kultur (360), 1934; *Dirigentenerfahrungen mit klassischen Meisterwerken (361), 1934; Anmerkungen zur Aufführungen von Beethoven Symphonien (362), ?1934

Über die Besetzung der Kurorchester (363), 1934; Brief anstelle eines Vorwortes zu Joseph Gregor, Richard Strauss|: Der Meister der Oper] (363A), 1935; Geschichte der Schweigsamen Frau (364), 1935; Arabella (365), 1937; *Bemerkungen zu Richard Wagners Gesamtkunstwerk und zum Bayreuther Festspielhaus (366), 1940; Bemerkungen zu Wagners Oper und Drama (367), ?1940; *Erinnerungen an meinen Vater (325), c1940; *Aus meinen Jugend- und Lehrjahren (326), c1940; *Vom melodischen Einfall (368), c1940; Omaggio a Giuseppe Verdi (368A), 1941; Anstelle eines Vorwortes zu Anton Berger: Richard Strauss als geistige Macht (369), 1941; *Zur Josephslegende (370), 1941; *Meine Werke in guter Zusammenstellung (371), 1941; Mozart, der Dramatiker (373), 1941; Glückwunsch für die Wiener Philharmoniker (374), 1942; *Geleitwort zu Capriccio (375), 1942; *Erinnerungen an die ersten Aufführungen meiner Opern von Guntram bis Intermezzo (376), 1942; Vorwort zu Divertimento (377), 1942; Über Wesen und Bedeutung der Oper (*rev. version) (378, 385), 1943, 1945; Geleitwort zu Willi Schuh: Das Bühnenwerk von Richard Strauss (379), 1944; *Über Mozart (380), 1944; Zum Kapitel Mozart (381), 1944; Über die Generalprobe der Oper Die Liebe der Danae (382), 1944; Gedanken über die Weltgeschichte des Theaters und Entwurf eines Briefes an Josef Gregor, den Verfasser (383), 1945; *Brief über das humanistische Gymnasium an Professor Reisinger (384), 1945; Geschichte der Oper Die Liebe der Danae (386), 1946; Pauline Strauss-de Ahna (387), 1847; *Glückwunsch für die sächsische Staatskapelle (388), 1948; Garmischer Rede am 85. Geburtstag (389), 1949; Letzte Aufzeichnung (390), 1949

Bibliography

BIBLIOGRAPHY

CORRESPONDENCE

F. Strauss, ed.: *Richard Strauss: Briefwechsel mit Hugo von Hofmannsthal* (Berlin, 1925; Eng. trans., 1928)

A. Mann, ed.: 'The Artistic Testament of Richard Strauss', *MQ*, xxxvi (1950), 1

Richard Strauss et Romain Rolland: correspondance, fragments de journal (Paris, 1951; Eng. trans., 1968)

F. von Schuch: *Richard Strauss, Ernst von Schuch und Dresdens Oper* (Leipzig, 1952, 2/1953)

W. Schuh, ed.: *Richard Strauss und Hugo von Hofmannsthal: Briefwechsel: Gesamtausgabe* (Zurich, 1952, enlarged 2/1955; Eng. trans., 1961/R1980)

W. Schuh and F. Trenner, eds.: 'Hans von Bülow/Richard Strauss: Briefwechsel', *Richard Strauss Jb 1954*, 7–88; separate Eng. trans. (1955)

W. Schuh, ed.: *Richard Strauss: Briefe an die Eltern 1882–1906* (Zurich, 1954)

R. Tenschert, ed.: *Richard Strauss und Joseph Gregor: Briefwechsel 1934–1949* (Salzburg, 1955)

F. Zagiba: 'Bella als Vorkämpfer des jungen Richard Strauss: Strauss' Künstlerisches Credo in sienen Briefen an Bella', *Johann Leopold Bella (1843–1936) und das Weiner Musikleben* (Vienna, 1955), 46

W. Schuh, ed.: *Richard Strauss, Stefan Zweig: Briefwechsel* (Frankfurt am Main, 1957; Eng. trans., 1977)

A. Holde: 'Unbekannte Briefe und Lieder von Richard Strauss', *NZM*, Jg.119 (1958), 71

E. Krause, ed.: 'Richard Strauss: ein Brief an Dora Wihan-Weis', *Richard Strauss Jb 1959–60*, 55

W. Schuh, ed.: 'Richard Strauss und Anton Kippenberg: Briefwechsel', *Richard Strauss Jb 1959–60*, 114–46

E. von Asow: 'Richard Strauss und Giuseppe Verdi', *ÖMz*, xvi (1961), 348

H. Fähnrich: 'Richard Strauss über das Verhältnis von Dichtung und Musik (Wort und Ton) in seinem Opernschaffen', *Mf*, xiv (1961), 22

'Erlebnis und Bekenntnis des jungen Richard Strauss', *Internationale Mitteilungen: Richard-Strauss-Gesellschaft* (1961), no.30, p.1 [letter to Cosima Wagner]

W. Schmieder: '57 unveröffentlichte Briefe und Karten von Richard Strauss in der Stadt- und Universitätsbibliothek Frankfurt/Main', *Festschrift Helmuth Osthoff* (Tutzing, 1961), 163

D. Kämper, ed.: *Richard Strauss und Franz Wüllner im Briefwechsel* (Cologne, 1963)

253

G. K. Kende and W. Schuh, eds.: *Richard Strauss, Clemens Krauss: Briefwechsel* (Munich, 1963, 2/1964)

W. Schuh, ed.: 'Richard Strauss, Briefe an Fritz Busch', *SMz*, civ (1964), 210

F. Grasberger: 'Gustav Mahler und Richard Strauss: aus Briefen und Tagebüchern', *ÖMz*, xxi (1966), 280

G. K. Kende: 'Aus dem Briefwechsel Richard Strauss/Clemens Krauss: 12 unveröffentlichte Briefe', *SMz*, cvi (1966), 2

F. Grasberger, ed.: *Der Strom der Töne trug mich fort: die Welt um Richard Strauss in Briefen* (Tutzing, 1967)

A. Ott: 'Richard Strauss und sein Verlegerfreund Eugen Spitzweg', *Musik und Verlag: Karl Vötterle zum 65. Geburtstag* (Kassel, 1968), 466

——, ed.: *Richard Strauss und Ludwig Thuille: Briefe der Freundschaft 1877–1907* (Munich, 1969)

W. Schuh, ed.: *Richard Strauss: Briefwechsel mit Willi Schuh* (Zurich, 1969)

M. Lemaire, ed.: 'Lettres inédites de Richard Strauss à Sylvain Dupuis', *Revue générale: perspectives européenes des sciences humaines*, viii (1970), 33

K. W. Birkin: 'Strauss, Zweig and Gregor: Unpublished Letters', *ML*, lvi (1975), 180

G. Brosche: 'Richard Strauss und Ludwig Karpath', *ÖMz*, xxxii (1977), 74

F. Trenner, ed.: *Cosima Wagner–Richard Strauss: ein Briefwechsel* (Tutzing, 1978)

H. Blaukopf, ed.: *Gustav Mahler, Richard Strauss: Briefwechsel 1888–1911* (Munich and Zurich, 1980)

F. Trenner, ed.: *Richard Strauss–Ludwig Thuille: ein Briefwechsel* (Tutzing, 1980)

A. Strauss, ed.: 'Richard Strauss–Manfred Mautner Markhof: Briefwechsel', *Richard-Strauss-Blätter*, new ser., no.5 (1981), 5

D. Wünsche, ed.: 'Richard Strauss und Heinz Tiessen: Briefwechsel', *Richard-Strauss-Blätter*, new. ser., no.6 (1981), 23

F. Hörburger: 'Über einige Briefe von Richard Strauss an Franz Carl Hörburger', *Gedenkschrift Hermann Beck* (Laaber, 1982), 201

G. Brosche, ed.: *Richard Strauss–Franz Schalk: ein Briefwechsel* (Tutzing, 1983)

D. Wünsche, ed.: 'Gerhart Hauptmann–Richard Strauss: Briefwechsel', *Richard-Strauss-Blätter*, new ser., no.9 (1983), 3–39

CATALOGUES, DOCUMENTARY MATERIAL, MEMOIRS, ICONOGRAPHY

R. Specht: *Richard Strauss: vollständiges Verzeichnis der im Druck erschienenen Werke* (Vienna, 1910)

Bibliography

T. Schäfer: *Also sprach Richard Strauss zu mir: aus dem Tagebuch eines Musikers und Schriftstellers* (Dortmund, 1924)

E. Wachten: *Richard Strauss, geboren 1864: sein Leben in Bildern* (Leipzig, 1940)

S. Zweig: *Die Welt von gestern: Erinnerungen eines Europäers* [1942] (Stockholm, 1944; Eng. trans., 1943)

R. Tenschert: *Anekdoten um Richard Strauss* (Vienna, 1945)

F. Busch: *Aus dem Leben eines Musikers* (Zurich, 1949)

E. Roth, ed. *Richard Strauss: Bühnenwerke* (London, 1954) [text in Ger., Eng., Fr.]

W. Schuh: *Das Bühnenwerk von Richard Strauss in den unter Mitwirkung des Komponisten geschaffenen letzten Münchner Inszenierungen* (Zurich, 1954)

F. Trenner: *Richard Strauss: Dokumente seines Lebens und Schaffens* (Munich, 1954)

L. Kusche: *Heimliche Aufforderung zu Richard Strauss mit 15 zeitgenössischen Karikaturen* (Munich, 1959)

R. Petzoldt: *Richard Strauss: sein Leben in Bildern* (Leipzig, 1962)

H. Zurlinden, ed.: *Erinnerungen an Richard Strauss: Carl Spitteler, Albert Schweitzer, Max Huber, Cuno Amiet, Artur Honegger* (St Gall, 1962)

A. Jefferson: *The Operas of Richard Strauss in Britain 1910–1963* (London, 1963)

B. Domin: *Richard Strauss in Würdigung seines 100. Geburtstages am 11. Juni 1964: eine Auswahl aus den Beständen der Stadtbibliothek Koblenz, Musikbücherei* (Koblenz, 1964)

F. E. Dostal, ed.: *Karl Böhm: Begegnung mit Richard Strauss* (Vienna, 1964)

F. Grasberger and F. Hadamowsky, eds.: *Richard-Strauss-Ausstellung zum 100. Geburtstag* (Vienna, 1964)

L. Kusche: *Richard Strauss im Kulturkarussell der Zeit 1864–1964* (Munich, 1964)

O. Ortner, ed.: *Richard-Strauss-Bibliographie, Teil 1: 1882–1944* (Vienna, 1964)

A. Ott, ed.: *Richard Strauss-Festjahr München 1964 zum 100. Geburtstag von Richard Strauss* (Munich, 1964)

W. Schuh: 'Die Richard-Strauss-Ausstellungen in Wien und München', *SMz*, civ (1964), 252

——: *Ein paar Erinnerungen an Richard Strauss* (Zurich, 1964)

W. Schuh and E. Roth, eds.: *Richard Strauss: Complete Catalogue* (London, 1964)

S. von Scanzoni, ed.: *Katalog der Ausstellung: Richard Strauss und seine Zeit* (Munich, 1964)

W. Thomas: *Richard Strauss und seine Zeitgenossen* (Munich, 1964)

F. Grasberger: *Richard Strauss: hohe Kunst, erfülltes Leben* (Vienna, 1965)

K. Böhm: *Ich erinnere mich ganz genau* (Zurich, 1968)

W. Deppisch: *Richard Strauss in Selbstzeugnissen und Bilddokumenten* (Reinbek, nr. Hamburg, 1968)

G. Brosche, ed.: *Richard-Strauss-Bibliographie, Teil 2: 1944–1964* (Vienna, 1973)

A. Jefferson: *Richard Strauss* (London, 1975)

A. Ott: 'Die Briefe von Richard Strauss in der Stadtbibliothek München', *Beiträge zur Musikdokumentation: Franz Grasberger zum 60. Geburtstag* (Tutzing, 1975), 341

F. Trenner: *Die Skizzenbücher von Richard Strauss aus dem Richard-Strauss-Archiv in Garmisch* (Tutzing, 1977)

G. Brosche and K. Dachs, eds.: *Richard Strauss: Autographen in München und Wien Verzeichnis* (Tutzing, 1979)

G. Jaaks and A. W. Jahnke, eds.: *Richard Strauss 1864–1949: Musik des Lichts in dunkler Zeit, vom Bürgerschreck Zum Rosenkavalier* (Mainz, 1980)

SPECIAL PERIODICAL ISSUES

Die Musik, iv/8 (1905)

Tempo (1949), no.12

Mitteilungen der Internationalen Richard-Strauss-Gesellschaft, i–lxiii (Berlin, 1952–69)

W. Schuh, ed.: *Richard Strauss Jb 1954*

———: *Richard Strauss Jb 1959–60*

Tempo (1964), no.69 [centenary number]

Richard-Strauss-Blätter, nos.1–12 (Vienna, 1971–8)

Richard-Strauss-Blätter, new ser. (Tutzing, 1979–)

LIFE, WORKS

G. Brecher: *Richard Strauss: eine monographische Skizze* (Leipzig, 1900)

R. Batka: *Richard Strauss* (Charlottenburg, 1908)

M. Steinitzer: *Richard Strauss* (Berlin, 1911, final enlarged edn. 1927)

H. T. Finck: *Richard Strauss: the Man and his Work* (Boston, 1917)

S. Kallenberg: *Richard Strauss: Leben und Werk* (Leipzig, 1926)

J. F. Cooke: *Richard Strauss: a Short Biography* (Philadelphia, 1929)

E. Gehring, ed.: *Richard Strauss und seine Vaterstadt: zum 70. Geburtstag am 11. Juni 1934* (Munich, 1934)

F. Gysi: *Richard Strauss* (Potsdam, 1934)

J. Kapp: *Richard Strauss und die Berliner Oper* (Berlin, 1934–9)

W. Brandl: *Richard Strauss: Leben und Werk* (Wiesbaden, 1949)

Bibliography

E. Bücken: *Richard Strauss* (Kevelaer, 1949)

K. Pfister: *Richard Strauss: Weg, Gestalt, Denkmal* (Vienna, 1949)

C. Rostand: *Richard Strauss* (Paris, 1949, 2/1965)

R. Tenschert: *Richard Strauss und Wien: eine Wahlverwandtschaft* (Vienna, 1949)

O. Erhardt: *Richard Strauss: Leben, Wirken, Schaffen* (Olten, 1953)

E. Krause: *Richard Strauss: Gestalt und Werk* (Leipzig, 1955, rev. 3/1963, rev. 6/1980; Eng. trans., 1964)

N. Del Mar: *Richard Strauss: a Critical Commentary on his Life and Works* (London, 1962–72/R with corrections 1978)

H. Kralik: *Richard Strauss: Weltbürger der Musik* (Vienna, 1963)

F. Hadamowsky: *Richard Strauss und Salzburg* (Salzburg, 1964)

W. Panofsky: *Richard Strauss: Partitur eines Lebens* (Munich, 1965)

G. R̄. Marek: *Richard Strauss: the Life of a Non-hero* (New York, 1967)

F. Grasberger: *Richard Strauss und die Wiener Oper* (Tutzing, 1969)

D. Jameux: *Richard Strauss* (Paris, 1971)

A. Jefferson: *The Life of Richard Strauss* (Newton Abbot, 1973)

M. Kennedy: *Richard Strauss* (London, 1976/R1983)

W. Schuh: *Richard Strauss: Jugend und Meisterjahre: Lebenschronik 1864–98* (Zurich, 1976; Eng. trans., 1982)

LIFE: ASPECTS AND EPISODES

J. C. Mannifarges: *Richard Strauss als Dirigent* (Amsterdam, 1907)

R. Müller-Hartmann: 'Reminiscences of Reger and Strauss', *ML*, xxix (1948), 153

R. Myers: 'Richard Strauss, 1864–1949', *MT*, xc (1949), 347

G. Samazeuilh: 'Richard Strauss and France', *Tempo* (1949), no.12, p.31

W. Schuh: 'In honour of Richard Strauss', *Tempo* (1949), no.12, p.5

E. Wellesz: 'Richard Strauss, 1864–1949', *MR*, xi (1950), 23

W. Reich: 'Richard Strauss und Romain Rolland', *Melos*, xviii (1951), 70

E. Roth: 'Richard Strauss in London 1947', *Richard Strauss Jb 1954*, 132

W. Schuh, ed.: 'Richard Strauss: Tagebuch der Griechenland- und Ägyptenreise (1892)', *Richard Strauss Jb 1954*, 89

R. Tenschert: 'Richard Strauss und die Salzburger Festspiele', *Richard Strauss Jb 1954*, 150

F. Trenner: 'Richard Strauss und Ernst von Wolzogen', *Richard Strauss Jb 1954*, 110

R. Tenschert: 'Richard Strauss und Mahler', *MJb 1954*, 195

J. Patzak: 'Richard Strauss als Mozart-Dirigent', *ÖMz*, xi (1956), 274

H. Fähnrich: 'Europäische Begegnung: Richard Strauss und Romain Rolland', *Musica*, xi (1957), 65

H. Friess: 'Richard Strauss and the Bavarian State Opera', *Tempo* (1957), no.43, p.26

G. Kende: 'Clemens Krauss über seine Zusammenarbeit mit Richard Strauss: nach Gesprächen aufgezeichnet', *SMz*, xcvii (1957), 4

H. Swarowsky: 'Persönliches von Richard Strauss', *ÖMz*, xii (1957), 137, 186

R. Tenschert: 'Richard Strauss: Gedenken der Rheinoper', *ÖMz*, xiv (1959), 421

J. von Rauchenberger-Strauss: 'Jugenderinnerungen', *Richard Strauss Jb 1959–60*, 7–30

G. K. Kende: *Richard Strauss und Clemens Krauss: eine Künstlerfreundschaft und ihre Zusammenarbeit an Capriccio (op.85): Konversationsstück für Musik* (Munich, 1960)

S. von Scanzoni: *Richard Strauss und seine Sänger: eine Plauderei über das Musiktheater in den Wind gesprochen* (Munich, 1961)

G. Samazeuilh: 'Richard Strauss as I Knew him', *Tempo* (1964), no.49, p.14

F. Trenner: 'Richard Strauss and Munich', *Tempo* (1964), no.49, p.5

L. Wurmser: 'Richard Strauss as an Opera Conductor', *ML*, xlv (1964), 4

T. Adorno: 'Richard Strauss. Born June 11, 1864', *PNM*, iv/1 (1965), 14; iv/2 (1966), 113

F. Strauss: 'Aphorismen um Richard Strauss', *ÖMz*, xxiii (1968), 336

R. Tenschert: 'Richard Strauss und Stefan Zweig', *ÖMz*, xxiii (1968), 75

F. Howes: 'Nimrod on Strauss', *MT*, cxi (1970), 590

J. Knaus: 'Leoš Janáček und Richard Strauss', *NZM*, Jg.133 (1972), 128

M. See: 'Richard Strauss und Romain Rolland: Bilanz einer Lebensfreundschaft', *NZM*, Jg.134 (1973), 204–72, 347, 413

R. Sterl: 'Das 2. Bayerische Musikfest 1904 unter Richard Strauss in Regensburg', *Mitteilungsblatt der Gesellschaft für Bayerische Musikgeschichte*, vi (1973), 97

W. Schuh: *Straussiana aus vier Jahrzehnten* (Tutzing, 1981)

P. Collins Jones: 'Richard Strauss, Ferruccio Busoni and Arnold Schönberg: "Some Imperfect Wagnerites" ', *MR*, xliii (1982), 169

P. M. Potter: 'Strauss's *Friedenstag*: a Pacifist Attempt at Political Resistance', *MQ*, lxix (1983), 408

WORKS: GENERAL

G. Jourissenne: *Richard Strauss: essai critique et biographique* (Brussels, 1899)

Bibliography

E. Urban: *Richard Strauss* (Berlin, 1901)

———: *Strauss contra Wagner* (Berlin, 1902)

A. Guttmann: 'Richard Strauss als Lyriker', *Die Musik*, iv (1904–5), 93

O. Bie: *Die moderne Musik und Richard Strauss* (Berlin, 1906)

P. Marsop: 'Italien und der "Fall Salome", nebst Glossen zur Kritik und Ästhetik', *Die Musik*, vi (1906–7), 139

E. Newman: *Richard Strauss* (London, 1908/*R*1970)

F. Santoliquido: *Il dopo-Wagner: Claude Debussy e Richard Strauss* (Rome, 1909)

C. Paglia: *Strauss, Debussy e compagnia bella: saggio di critica semplicista e spregiudicata per il gran pubblico* (Bologna, 1913)

A. Seidl: *Straussiana: Aufsätze zur Richard Strauss-Frage aus drei Jahrzehnten* (Regensburg, 1913)

O. Bie: *Die neuere Musik bis Richard Strauss* (Leipzig, 1916)

P. Bekker: *Kritische Zeitbilder* (Stuttgart, 1921)

R. Specht: *Richard Strauss und sein Werk* (Leipzig, 1921)

H. W. von Waltershausen: *Richard Strauss: ein Versuch* (Munich, 1921)

E. Bloch: 'Mahler, Strauss, Bruckner', *Die Musik*, xv (1922–3), 664

W. Klatte: 'Aus Richard Strauss' Werkstatt', *Die Musik*, xvi (1923–4), 636

M. Steinitzer: 'Der unbekannte Strauss', *Die Musik*, xvi (1923–4), 653

H. Windt: 'Richard Strauss und die Atonalität', *Die Musik*, xvi (1923–4), 642

W. Schrenk: *Richard Strauss und die neue Musik* (Berlin, 1924)

R. Tenschert: 'Die Kadenzbehandlung bei Richard Strauss: ein Beitrag zur neueren Harmonik', *ZMw*, viii (1925), 161

A. Seidl: *Neuzeitliche Tondichter und zeitgenössische Tonkünstler: gesammelte Aufsätze, Studien und Skizzen* (Regensburg, 1926)

E. Stein: 'Mahler, Reger, Strauss und Schönberg: kompositionstechnische Betrachtungen', *25 Jahre neue Musik: Jb 1926 der Universal-Edition* (Vienna, 1926), 63

P. Bekker: 'Brief an Richard Strauss', *Die Musik*, xxv (1932–3), 81

R. Tenschert: 'Die Tonsymbolik bei Richard Strauss', *Die Musik*, xxvi (1933–4), 646

W. Reich: 'Bemerkungen zum Strauss'schen Opernschaffen, anlässlich des 70. Geburtstages', *Der Auftakt*, xiv (1934), 101

R. Tenschert: 'Versuch einer Typologie der Richard Strauss'schen Melodik', *ZMw*, xvi (1934), 274

———: 'Autobiographisches im Schaffen von Richard Strauss', *ZfM*, Jg.106 (1939), 582

———: *Dreimal sieben Variationen über das Thema Richard Strauss* (Vienna, 1944, 2/1945)

W. Schuh: 'Unvollendete Spätwerke von Richard Strauss', *SMz*, xc (1950), 392

R. Tenschert: 'Richard Strauss-Uraufführung', *ZfM*, cxiii (1952), 248

H. Osthoff: 'Mozarts Einfluss auf Richard Strauss', *SMz*, xcviii (1958), 409

E. Newman: *Testament of Music* (London, 1963)

A. Berger: *Richard Strauss als geistige Macht: Versuch eines philosophischen Verständnisses* (Gisch, 1964)

F. Hadamowsky: 'Die Wiener Ur- und Erstaufführungen der Werke von Richard Strauss', *ÖMz*, xix (1964), 233

R. Strauss: ' "Über mein Schaffen": eine bisher unveröffentliche Skizze', *ÖMz*, xix (1964), 221

L. Wurmser: 'Strauss and some Classical Symphonies', *ML*, xlv (1964), 233

A. Goléa: *Richard Strauss* (Paris, 1965)

R. Gerlach: 'Richard Strauss: Prinzipien seiner Kompositionstechnik', *AMw*, xxiii (1966), 277

——: *Tonalität und tonale Konfiguration im Oeuvre von Richard Strauss: Analysen und Interpretationen als Beiträge zum Verständnis von tonalen Problemen und Formen in sinfonischen Werken und in der 'Einleitung' und ersten Szene des Rosenkavalier* (diss., U. of Zurich, 1966; Berne, 1966, as *Don Juan und Rosenkavalier*)

E. Roth: *Musik als Kunst und Ware: Betrachtungen und Begegnungen eines Musikverlegers* (Zurich, 1966)

Festschrift Dr. Franz Strauss (Tutzing, 1967)

A. Ott: 'Documenta musicae domesticae Straussiana', *Festschrift Dr. Franz Strauss* (Tutzing, 1967), 65

E. Roth: 'Das Königliche in der Musik von Richard Strauss', *Festschrift Dr. Franz Strauss* (Tutzing, 1967), 91

F. Trenner: 'Richard Strauss und seine Vaterstadt', *Festschrift Dr. Franz Strauss* (Tutzing, 1967), 137

W.-E. von Lewenski: 'Die Rolle der Klangfarbe bei Strauss und Debussy', *SMz*, cx (1970), 357

A. A. Abert: 'Richard Strauss und das Erbe Wagners', *Mf*, xxvii (1974), 165

A. Forchert: 'Zur Auflösung traditioneller Formkategorien in der Musik um 1900: Probleme formaler Organisation bei Mahler und Strauss', *AMw*, xxxii (1975), 85

A. Batta: 'A Nietzsche Symbol in the Music of Richard Strauss and Béla Bartók', *New Hungarian Quarterly*, xxii (1982), 202

OPERAS

L. Gilman: *Strauss' Salome: a Guide to the Opera* (New York, 1906)

E. Schmitz: *Richard Strauss als Musikdramatiker: eine aesthetisch-kritische Studie* (Munich, 1907)

Bibliography

P. Bekker: *Das Musikdrama der Gegenwart* (Stuttgart, 1909)

———: '*Elektra*: Studie', *Neue Musik-Zeitung*, xxx (1909), 293, 330, 387

E. Fischer-Plasser: *Einführung in die Musik von Richard Strauss und Elektra* (Leipzig, 1909)

G. Gräner: *Richard Strauss: Musikdramen* (Berlin, 1909)

C. Mennicke: 'Richard Strauss: *Elektra*', *Riemann-Festschrift* (Leipzig, 1909), 503

F. Torrefranca: 'Riccardo Strauss e l'*Elektra*', *RMI*, xvi (1909), 335–84

O. R. Hübner: *Richard Strauss und das Musikdrama: Betrachtungen über den Wert oder Unwert gewisser Opernmusiken* (Leipzig, 1910)

E. Hutcheson: *Elektra by Richard Strauss: a Guide to the Opera with Musical Examples from the Score* (New York, 1910)

W. Klein: 'Die Harmonisation in *Elektra* von Richard Strauss: ein Beitrag zur modernen Harmonisationslehre', *Der Merker*, ii (1911), 512, 540, 590

F. Torrefranca: 'Il Rosenkavalier di Riccardo Strauss', *RMI*, xviii (1911), 147–79

———: 'La nuova opera di Riccardo Strauss', *RMI*, xix (1912), 986–1031

M. Steinitzer: *Richard Strauss in seiner Zeit, mit einem Abdruck der auf der Strausswoche zu Stuttgart im Kgl. Hoftheater gehaltenen Rede und einem Bildnis* (Leipzig, 1914)

B. Diebold: 'Die ironische *Ariadne* und der *Bürger als Edelmann*', *Deutsche Bühne*, i (1918), 219

A. Rosenzweig: *Zur Entwicklungsgeschichte des Strauss'schen Musikdramas* (diss., U. of Vienna, 1923)

R. Specht: 'Vom *Guntram* zur *Frau ohne Schatten*', *Almanach der Deutschen Musikbücherei auf das Jahr 1923* (Regensburg, 1923), 150

K. Westphal: 'Das musikdramatische Prinzip bei Richard Strauss', *Die Musik*, xix (1926–7), 859

G. Röttger: *Die Harmonik in Richard Strauss' Der Rosenkavalier: ein Beitrag zur Entwicklung der romantischen Harmonik nach Richard Wagner* (diss., U. of Munich, 1931)

K. H. Ruppel: 'Richard Strauss und das Theater', *Melos*, xiii (1934), 175

K.-J. Krüger: *Hugo von Hofmannsthal und Richard Strauss: Versuch einer Deutung des künstlerischen Weges Hugo von Hofmannsthals, mit einem Anhang: erstmalige Veröffentlichung der bisher ungedruckten einzigen Vertonung eines Hofmannsthalschen Gedichtes durch Richard Strauss* (Berlin, 1935)

H. Röttger: *Das Formproblem bei Richard Strauss gezeigt an der Oper*

Die Frau ohne Schatten, mit Einschluss von Guntram und Intermezzo (diss., U. of Munich, 1935; abridged, Berlin, 1937)

J. Gregor: *Richard Strauss: der Meister der Oper* (Munich, 1939, 2/1942)

——: 'Zur Entstehung von Richard Strauss' *Daphne*', *Almanach zum 35. Jahr des Verlags R. Piper & Co., München* (Munich, 1939), 104

R. Tenschert: 'Hosenrollen in den Bühenwerken von Richard Strauss', *ZfM*, Jg.106 (1939), 586

R. Hartmann: *Capriccio: ein Konversationsstück für Musik in 1 Aufzug von Clemens Krauss und Richard Strauss, op.85: Regieangaben nach Erfahrungen der Uraufführung, Staatsoper München, 28. Oktober 1942* (Berlin, 1943)

G. Becker: *Das Problem der Oper an Hand von Richard Strauss' Capriccio* (diss., U. of Jena, 1944)

A. Mathis: 'Stefan Zweig as Librettist and Richard Strauss', *ML*, xxv (1944), 163, 226

W. Schuh: 'Eine nicht komponierte Szene zur *Arabella*', *SMz*, lxxxiv (1944), 231

O. Gatscha: *Librettist und Komponist: dargestellt an den Opern Richard Strauss'* (diss., U. of Vienna, 1947)

A. Pryce-Jones: *Richard Strauss: Der Rosenkavalier* (London, 1947)

W. Schuh: *Über Opern von Richard Strauss* (Zurich, 1947)

O. Erhardt: 'The Later Operatic Works of Richard Strauss', *Tempo* (1949), no.12, p.23

E. Roth: 'Don Juan: Some Letters about the First Performance', *Tempo* (1949), no.12, p.11

F. Trenner: *Die Zusammenarbeit von Hugo von Hofmannsthal und Richard Strauss* (diss., U. of Munich, 1949)

O. Erhardt: 'Richard Strauss's *Die Frau ohne Schatten*', *Tempo* (1950), no.17, p.24

S. Beinl: 'A Producer's Viewpoint: Notes on *Die Frau ohne Schatten*', *Tempo* (1952), no.24, p.29

D. Lindner: *Richard Strauss/Joseph Gregor: Die Liebe der Danae: Herkunft, Inhalt und Gestaltung eines Opernwerkes* (Vienna, 1952)

R. Schopenhauer: *Die antiken Frauengestalten bei Richard Strauss* (diss., U. of Vienna, 1952)

W. Schuh, ed.: *Hugo von Hofmannsthal: Danae oder die Vernunftheirat: Szenarium und Notizen* (Frankfurt, 1952)

R. Tenschert: 'A "gay myth": the Story of *Die Liebe der Danae*', *Tempo* (1952), no.24, p.5

E. Wellesz: 'Hofmannsthal and Strauss', *ML*, xxxiii (1952), 239

G. Hausswald: *Richard Strauss: ein Beitrag zur Dresdener Operngeschichte seit 1945* (Dresden, 1953)

E. Lismann: 'Ariadne auf Naxos', *Tempo* (1953), no.28, p.25

Bibliography

W. Wendhausen: *Das stilistische Verhältnis von Dichtung und Musik in der Entwicklung der musikdramatischen Werke Richard Strauss'* (diss., U. of Hamburg, 1954)

R. Tenschert: 'Composer and Librettist', *Tempo* (1956), no.41, p.24

——: '*Arabella*: die letzte Gemeinschaft von Hugo von Hofmannsthal und Richard Strauss', *ÖMz*, xiii (1958), 323

——: 'The Sonnet in Richard Strauss's Opera "Capriccio"', *Tempo* (1958), no.47, p.7

G. Baum: '*Hab' mir's gelobt, ihn lieb zu haben* . . .': *Richard Strauss und Hugo von Hofmannsthal nach ihrem Briefwechsel dargestellt* (Berlin, 1962)

A. A. Abert: 'Stefan Zweigs Bedeutung für das Alterswerk von Richard Strauss', *Festschrift Friedrich Blume zum 70. Geburtstag* (Kassel, 1963), 7

E. Graf: 'Die Bedeutung von Richard Strauss' *Intermezzo*', *ÖMz*, xviii (1963), 241

A. Natan: *Richard Strauss: die Opern* (Basle, 1963)

H. Redlich: ' "Prima la musica . . .?": a Ruminative Comment on Richard Strauss' Final Opera', *MR*, xxiv (1963), 185

L. Lehmann: *Five Operas and Richard Strauss* (New York, 1964; as *Singing with Richard Strauss*, London, 1964)

W. Mann: *Richard Strauss: a Critical Study of the Operas* (London, 1964)

K. Pörnbacher: *Hugo von Hofmannsthal/Richard Strauss: Der Rosenkavalier* (Munich, 1964)

W. Schuh: *Hugo von Hofmannsthal und Richard Strauss: Legende und Wirklichkeit* (Munich, 1964)

H. Grabner: 'Eine denkwürdige Erstauf-Erstaufführung', *40 Jahre steierischer Tonkünstlerbund* (Graz, 1967), 49 [*Salome*]

R. Wittelsbach: 'Betrachtungen zu *Salome* und *Elektra*', *Festschrift 1817–1967: Akademie für Musik und darstellende Kunst in Wien* (Vienna, 1967), 93

H. Schnoor: *Die Stunde des Rosenkavalier: 300 Jahre Dresdner Oper* (Munich, 1968)

W. Schuh: *Der Rosenkavalier: 4 Studien* (Olten, 1968)

H. Becker: 'Richard Strauss als Dramatiker', *Beiträge zur Geschichte der Oper* (Regensburg, 1969), 165

O. Schneider: '*Der Rosenkavalier* in Salzburg', *ÖMz*, xxiv (1969), 451

A. Wasdruszka: 'Das *Rosenkavalier* Libretto', *ÖMz*, xxiv (1969), 440

P. Burwik: *Die Bühnenbilder zu den Wiener Aufführungen von Richard Strauss-Opern, 1902–1964* (diss., U. of Vienna, 1970)

W. Schuh: 'Richard Strauss und seine Libretti', *GfMKB, Bonn 1970*, 169

263

J. Knaus: *Hugo von Hofmannsthal und sein Weg zur Oper Die Frau ohne Schatten* (Berlin, 1971)

W. Schuh, ed.: *Hugo von Hofmannsthal, Richard Strauss: Der Rosenkavalier: Fassungen, Filmszenarium, Briefe* (Frankfurt am Main, 1971)

A. A. Abert: *Richard Strauss: die Opern: Einführung und Analyse* (Hanover, 1972)

G. I. Ascher: *Die Zauberflöte und die Frau ohne Schatten: ein Vergleich zwischen zwei Operndichtungen der Humanität* (Berne, 1972)

A. A. Abert: 'Richard Strauss' Anteil an seinen Operntexten', *Musicae scientiae collectanea: Festschrift Karl Gustav Fellerer* (Cologne, 1973), 1

W. Keller: '*Die Liebe der Danae*: ein Wagner-Oper von Richard Strauss? Ein Beitrag zur ästhetischen Manipulation', *NZM*, Jg.134 (1973), 628

D. G. Daviau and G. J. Buelow: *The 'Ariadne auf Naxos' of Hugo von Hofmannsthal and Richard Strauss* (Chapel Hill, 1975)

R. Gerlach: 'Die ästhetische Sprache als Problem im Rosenkavalier', *NZM*, i (1975), 95, 278

W. Schuh: 'Hofmannsthals Randnotizen für Richard Strauss im *Ariadne* Libretto', *Für Rudolf Hirsch: zum siebzigsten Geburtstag* (Frankfurt am Main, 1975), 224

———: 'Metamorphosen einer Ariette von Richard Strauss', *Opernstudien: Anna Amalie Abert zum 65. Geburtstag* (Tutzing, 1975), 197

C. Höslinger: '*Salome* und ihr österreichisches Schicksal, 1905–1919', *ÖMz*, xxxii (1977), 300

K. Overhoff: *Die Elektra-Partitur von Richard Strauss* (Salzburg, 1978)

S. Pantle: '*Die Frau ohne Schatten*' by Hugo von Hofmannsthal and Richard Strauss: an Analysis of Text, Music and their Relationship* (Berne, 1978)

M. Enix: 'A Reassessment of *Elektra* by Strauss', *Indiana Theory Review*, ii (1979), 31

D. Zimmerschied: 'Integration in Liebe oder brutale Vertreibung? Versuche zur Deutung der Sängerepisode im "Rosenkavalier" ', *Mf*, xxxii (1979), 291

C. Erwin: 'Richard Strauss's Presketch Planning for *Ariadne auf Naxos*', *MQ*, lxvii (1981), 348

B. Gilliam: 'Stefan Zweig's Contribution to Strauss Opera after *Die schweigsame Frau*', *Richard-Strauss-Blätter*, new ser., no.5 (1981), 48

R. Hartmann: *Die Bühnenwerke von der Uraufführung bis heute* (Fribourg, 1980; Eng. trans., 1981)

Bibliography

K. Forsyth: *'Ariadne auf Naxos' by Hugo von Hofmannsthal and Richard Strauss: its Genesis and Meaning* (London and Oxford, 1982)

K. W. Birkin: 'Collaboration out of Crisis (Strauss–Zweig–Gregor)', *Richard-Strauss-Blätter*, new ser., no.9, ix (1983), 50

N. John, ed.: *Der Rosenkavalier*, English National Opera Guide no.8 (London, 1981) [incl. articles by D. Puffett, M. Kennedy, P. Branscombe, libs in Ger. and Eng.]

R. Sussman Stewart, ed.: *Der Rosenkavalier*, Metropolitan Opera Classics Library no.1 (London, 1983) [incl. articles by G. R. Marek, A. Burgess, libs in Ger. and Eng.]

ORCHESTRAL WORKS

G. Brecher: 'Richard Strauss als Symphoniker', *Leipziger Kunst*, i (1899), 400, 417

H. Merian: *Richard Strauss' Tondichtung 'Also sprach Zarathustra': eine Studie über die moderne Programmsymphonie* (Leipzig, 1899)

E. von Ziegler: *Richard Strauss und seine dramatischen Dichtungen* (Munich, 1907)

H. Walden, ed.: *Richard Strauss: Symphonien und Tondichtungen erläutert von G. Brecher, A. Hahn, W. Klatte, W. Mauke, A. Schattmann, H. Treibler, H. Walden, nebst einer Einleitung: Richard Strauss' Leben und Schaffen* (Berlin, 1908)

L. Gilman: 'Richard Strauss and his Alpine Symphony', *North American Review*, xxiv (1919), 920

H. Scherchen: 'Tonalitätsprinzip und die Alpen-Symphonie von Richard Strauss', *Melos*, i (1920), 198, 244

A. Lorenz: 'Der formale Schwung in Richard Strauss' *Till Eulenspiegel*', *Die Musik*, xvii (1924–5), 658

T. Armstrong: *Strauss' Tone Poems* (London, 1931)

E. Wachten: *Das psychotechnische Formproblem in den Sinfonischen Dichtungen von Richard Strauss* (*mit besonderer Berücksichtigung seiner Bühnenwerke*) (diss., U. of Berlin, 1932; abridged, Berlin, 1933)

R. Tenschert: 'Wandlungen einer Kadenz: Absonderlichkeiten der Harmonik im *Don Quixote* von Richard Strauss', *Die Musik*, xxvi (1933–4), 663

A. Lorenz: 'Neue Formerkenntnisse, angewandt auf Richard Straussens *Don Juan*', *AMf*, i (1936), 452

C. Blessing: *Das instrumentale Schaffen von Richard Strauss im Spiegelbild der Presse und der zeitgenössischen Kritik* (diss., U. of Munich, 1944)

R. Raffalt: *Über die Problematik der Programmusik: ein Versuch ihres Aufweises an der Pastoral-Symphonie von Beethoven, der Berg-*

Symphonie von Liszt und der Alpensinfonie von Strauss (diss., U. of Tübingen, 1949)

L. Kusche and K. Wilhelm: 'Richard Strauss's "Metamorphosen"', *Tempo* (1951), no.19, p.19

W. Brennecke: 'Die "Metamorphosen"-Werke von Richard Strauss und Paul Hindemith', *SMz*, ciii (1963), 129

R. Longyear: 'Schiller, Moszkowski, and Strauss: Joan of Arc's "Death and Transfiguration"', *MR*, xxviii (1967), 209

W. Gruhn: *Die Instrumentation in den Orchesterwerken von Richard Strauss* (diss., U. of Mainz, 1968)

K. Niemöller: 'Zur Ästhetik der Symphonischen Dichtungen von Richard Strauss', *Mitteilungen der Internationalen Richard-Strauss-Gesellschaft*, lvii–lix (1968), 3

P. Franklin: 'Strauss and Nietzsche: a Revaluation of "Zarathustra"', *MR*, xxxii (1971), 248

P. Damm: 'Gedanken zu den Hornkonzerten von Richard Strauss', *Richard-Strauss-Blätter*, new ser., no.4 (1980), 31

R. Schloetterer-Traimer: 'Béla Bartók und die Tondichtungen von Richard Strauss', *ÖMz*, xxxvi (1981), 311

B. Gellermann: '*Die Donau*: Betrachtungen zum Fragment der letzten Symphonischen Dichtung von Richard Strauss', *Richard-Strauss-Blätter*, new ser., no.7 (1982), 21

OTHER WORKS

F. Dubitzky: 'Richard Strauss' Kammermusik', *Die Musik*, xiii (1913–14), 283

E. Thilo: 'Richard Strauss als Chorkomponist', *Die Musik*, xiii (1913–14), 304

M. Steinitzer: 'Richard Strauss' Werke für Klavier', *Die Musik*, xxiv (1931–2), 105

R. Tenschert: 'Verhältnis von Wort und Ton: eine Untersuchung an dem Strauss'schen Lied "Ich trage meine Minne"', *ZfM*, Jg.101 (1934), 591

W. Schuh: 'Ein vergessenes Goethe-Lied von Richard Strauss', *SMz*, lxxix (1949), 235

A. Hutchings: 'Strauss's Four Last Songs', *MT*, xci (1950), 465

W. Schuh: 'Richard Strauss's "Four Last Songs"', *Tempo* (1950), no.15, p.25

A. Orel: 'Richard Strauss als Begleiter seiner Lieder', *SMz*, xcii (1952), 12

R. Breuer: 'Drei "neue" Lieder von Richard Strauss', *SMz*, xcix (1959), 10

H. Friess: 'Eine Bühnenmusik zu Shakespeares *Romeo und Julia*', *Richard Strauss Jb 1959–60*, 51–83

Bibliography

W. Schuh: 'Das Szenarium und die musikalischen Skizzen zum Ballet *Kythere*', *Richard Strauss Jb 1959–60*, 84

R. Tenschert: 'Der "Krämerspiegel" von Richard Strauss', *ÖMz*, xvi (1961), 221

R. H. Schäfer: *Hugo von Hofmannsthals Arabella* (Berne, 1967)

H. Federhofer: 'Die musikalische Gestaltung des *Krämerspiegels* von Richard Strauss', *Musik und Verlag: Karl Vötterle zum 65. Geburtstag* (Kassel, 1968), 260

G. Baum: 'Hugo Wolf und Richard Strauss in ihren Liedern', *NZM*, Jg.130 (1969), 575

E. Lockspeiser: 'The Berlioz–Strauss Treatise on Instrumentation', *ML*, l (1969), 37

A. Jefferson: *The Lieder of Richard Strauss* (London, 1971)

W. W. Colson: *Four Last Songs by Richard Strauss* (diss., U. of Illinois, 1974)

U. Lienenlüke: *Lieder von Richard Strauss nach zeitgenössischer Lyrik* (Regensburg, 1976)

B. A. Petersen: *Ton und Wort: the Lieder of Richard Strauss* (Ann Arbor, 1980)

JEAN SIBELIUS

Robert Layton

CHAPTER ONE

Student years

Jean (Julius Christian) Sibelius was born on 8 December 1865 in Hämeenlinna (Tavastehus), a small town in south central Finland. His first name was Johan but he took its gallicized form in emulation of an uncle. He was the second of three children of a Swedish-speaking family; the others, an older sister (Linda) and a younger brother (Christian), both showed some musical aptitude. His father, a doctor, died during the cholera epidemic of 1867–8, and the family was brought up by the mother and grandmother. At the age of 11 Sibelius entered the first Finnish-speaking grammar school, the Hämeenlinna Suomalainen Normaalilyseo, and though he spoke some Finnish from the age of eight onwards, he did not acquire complete proficiency in the language until he was a young man. He showed an early musical ability both as a violinist and as a composer; his first effort at composition, a simple piece for violin and cello (pizzicato) called *Vattendroppar* ('Water drops'), comes from his tenth year. However, it was not until he was 14 that he began studying the violin in earnest with the local bandmaster Gustav Levander, and his developing prowess on the instrument encouraged his ambitions to be a soloist. He was an active chamber music player throughout these years (the family formed a trio), acquiring a sound and thorough knowledge of the Viennese classics, and continued to show an active interest in composition. He studied Marx's composition treatise and wrote a good deal of chamber music including the Piano Trio in A minor (1881–2) and Piano Quartet in E minor. During his school years he developed a keen interest in the *Kalevala*, the Finnish national epic, an interest which in his 20s was to deepen and become a vital source of inspiration. His innate and passionate love of nature drew him to the great Swedish-speaking nature poets, particularly Runeberg, many of whose poems he was to set in later life.

In 1885 Sibelius enrolled at the University of Helsinki in the faculty of law, but, like so many composers, he gave greater priority to music than his law studies. After a year music claimed him completely and he turned his energies to it, studying composition with Martin Wegelius and the violin with Hermann Czillag and Mitrofan Wasilyeff. He played the second violin in the string quartet which Wegelius's music school boasted, and as a soloist played Bériot's Violin Concerto no.7, Vieuxtemps' *Fantaisie-caprice*, the second and third movements of the Mendelssohn concerto, as well as the E minor concerto of Félicien David. Even though the violin consumed both his energies and ambitions, it became obvious by the late 1880s that his true path was that of a composer. But he still cherished plans as an executant and even as late as 1890–91, while living in Vienna, he auditioned for the Philharmonic. For all this, however, he never proceeded to the last part of the conservatory course, which would have entailed mastery of the Bach solo sonatas and the Paganini caprices.

In Helsinki (1885–9) the dominating influence on his musical development was Wegelius, a pupil of Reinecke, Hans Richter and Svendsen. A man of unusual ver-

25. *Jean Sibelius: detail from portrait (1894) by Akseli Gallén-Kallela in the Sibelius Museum, Ainola*

273

satility and wide experience, active as composer, pianist
and conductor, he was not slow to recognize Sibelius's
gifts, and took him under his wing. He gave him a good
grounding in harmony, counterpoint and fugue, though
less thorough and demanding than Sibelius later
received from Albert Becker in Berlin. Wegelius was a
keen Wagnerian but had relatively little sympathy for
Tchaikovsky, whose influence on Sibelius was strong at
this time. Sibelius made less contact with the orchestral
repertory than might have been expected during these
student years, as Wegelius discouraged his pupils from
attending Kajanus's symphony concerts: a bitter rivalry
existed between the two musicians, and it was not until
1890, when Kajanus visited Berlin to conduct his *Aino*
symphony, that Sibelius's long and fruitful friendship
with the conductor was put on a sure footing. Sibelius
did, however, derive enormous stimulus from his friendship with Busoni, who taught in Helsinki at Wegelius's
school and was a close companion.

Sibelius went to study abroad in 1889. At first he
had entertained the idea of continuing his studies at St
Petersburg, which was close to hand, with Rimsky-
Korsakov. However, Wegelius was against that and in
the autumn Sibelius was dispatched to Berlin to study
with Albert Becker. A strict contrapuntist of the old
school, Becker subjected his pupil to a rigorous and no
doubt chastening discipline. But if Sibelius complained
of Becker's pedantry ('he won't hear about anything
other than his fugues. It is deadly boring to concern
oneself exclusively with that. I know the German
psalmbook inside out'), and found Berlin itself un-
congenial, he was brought into contact with music-
making of a high order: he heard *Tannhäuser*, *Meister-*

singer, Bülow recitals of Beethoven sonatas and the Joachim Quartet's of late Beethoven, and attended the first performance of Strauss's *Don Juan*. He went with Busoni to Leipzig, where Busoni played in the Sinding E minor Piano Quintet and he worked on a piano quintet of his own. Generally speaking, however, this was a fallow period and apart from the quintet, the only work of note that he composed was the String Quartet in B♭ op.4.

Sibelius's financial extravagance and heavy drinking, characteristics shared with his father, caused problems in Berlin and subsequently in Vienna, where he went for a second year of foreign study. Busoni had supplied him with a letter of introduction to Brahms, but the latter refused to receive him and he eventually studied in Vienna with Karl Goldmark, then at the height of his fame, and Robert Fuchs. So far Sibelius had confined his composition to chamber music; it was in Vienna that he approached the orchestra. The main idea for *Kullervo* comes from this period, and he spoke to his biographer Ekman of an octet from which, it was thought, some of the themes of *En saga* were drawn. It is equally likely, however, that the octet was absorbed into music written to accompany a fencing match. In Vienna Sibelius was a frequent visitor at Pauline Lucca's salon, and it was there (and in Berlin) that he formed a taste for high society which ill accorded with his resources.

CHAPTER TWO

Early success

After Sibelius returned to Finland, in 1891, he was able to maintain himself by a certain amount of teaching, but the bulk of his energies at the end of that year and in the winter of 1892 were absorbed by *Kullervo*, which received its première in April. Its success was such that his position in Finnish musical life was never seriously challenged from that moment onwards. *Kullervo* was the first of Sibelius's many works to draw on the *Kalevala*. Considering that Sibelius's working knowledge of the orchestra was of such recent provenance, *Kullervo* shows astonishing assurance, and in style the work breaks entirely new ground in his output. The very opening bars proclaim that this is a new voice, and though the seams in its structure are clearly audible, the work leaves no doubt that its composer has a strong grasp of form and a sense of forward movement. It is a five-movement symphony of Mahlerian proportions. The vocal writing in the long centrepiece, 'Kullervo and his Sister', is both highly original and keenly dramatic; indeed it suggests that Sibelius could have developed an extremely respectable operatic style had he chosen to do so.

His marriage in June 1892 to Aino Järnefelt brought Sibelius into one of the most influential nationalist-liberal families in Finland. General Alexander Järnefelt was a keen advocate of the Finnish language, a sensitive political issue in the 1890s and for much of the early 20th century. Sibelius himself had the strongest national consciousness, and much of his music in the 1890s became a focus for patriotic sentiment at a time when Finland's aspirations to independence were stronger than ever and its autonomy within the Russian Empire was being slowly eroded. Immediately following the success of *Kullervo*, there followed a number of purely orchestral works, including *En saga*, which was written in response to a commission for Kajanus, and the *Karelia* music. However, by far the most powerful of his works before (and perhaps even including) the First Symphony are the four *Lemminkäinen* legends. *Tuonelan joutsen* ('The swan of Tuonela') was originally conceived as the prelude to the opera *Veenen luominen* ('The building of the boat') and though its original sound world is not in question, *Lohengrin* is obviously a distant relative. There is no doubt that some of the thematic substance of *Lemminkäinen Tuonelassa* ('Lemminkäinen in Tuonela') has operatic origins. The magical passage for strings in A minor (bar 186) turns up in the sketches for *The Building of the Boat*. Sibelius's visit to Bayreuth in summer 1894 confronted him with the realities of the operatic composer after Wagner. Indeed, in later years he was less than frank in explaining the impact Wagner had made on him during this visit. Far from leaving him indifferent, Wagner bowled him over and he attended more performances than he chose to tell his biographer, Ekman. Wagner's achievement almost certainly deterred him from embarking on an operatic course; he abandoned *The Building of the Boat* and apart from one other attempt, *Jungfrun i tornet* ('The maiden in the tower'), which was also hampered by a miserable libretto, turned his back on the genre. (At one time, he toyed with the notion of a comic opera to a libretto by Adolf Paul and Birger Mörner but this plan never came to anything.)

In spite of the fact that the *Lemminkäinen* legends did not enjoy a successful première (Flodin, who had championed Sibelius's work up to this point, regretted his increasing reliance on programmatic inspiration), they remain in their definitive form the most profoundly original, in both language and formal design, of Sibelius's works of the 1890s. He withheld two from publication until late in life for no reason other than the fact that Kajanus showed little enthusiasm for them. Their complete emergence substantiated at least to some extent their symphonic claims. Indeed, in his last years Sibelius referred to them as a symphony.

Like so many of his works they were subjected to thorough-going revision. This was a feature of Sibelius's creative personality: *En saga* was completely over-hauled in 1902, and the revision shows the extent to which his orchestral skill and formal awareness had developed in the intervening decade. He felt sufficiently sure of himself, for example, to elongate pedal points in the second version rather than contract them. He reduced the number of key changes from 48 in the first version to 34 in the second. As far as the orchestra is concerned, his imagination became much more vividly realized: the *divisi* strings at the opening are more effec-tively spaced, and transitions are linked with greater assurance. The legends underwent much the same process, and it was in the last of them, *Lemminkäisen paluu* ('Lemminkäinen's return'), that his symphonic mastery was first displayed. Although a strong sense of movement and feeling for form are evident in the first movement of *Kullervo* (and to a lesser extent in the first version of *En saga*), it is only in *Lemminkäinen's Return* that important aspects of the genuinely symphonic Sibelius appear: the ability to control momentum and the subtle transformation of thematic material to serve illustrative purposes show him operating with brilliance in both programmatic and symphonic dimensions.

The 1890s saw the formation and then consolidation of a personal language. Sibelius's early works for various chamber combinations show him responding in turn to the Viennese classics (the String Quartet in E♭ of 1885 is very Haydnesque), to Grieg and Svendsen (the F major Violin Sonata, written in 1889 according to the convincing argument of Rosas, is obviously indebted to Grieg, as is the slow movement of the String Quartet in B♭ op.4 of 1890) and above all to Tchaikovsky, particularly his harmonic vocabulary. This can readily be discerned in the student works (the String Trio in G minor of 1885) but is even more striking in the slow movement of *Kullervo*. It may be said to reach its climax in the slow movement of the First Symphony (1899), written in the wake, as it were, of Tchaikovsky's *Pathétique*, which was heard in Helsinki in 1894 and 1897. Wagner's influence can also be detected, powerfully in such songs – almost scenas – as *Höstkväll* ('Autumn evening') or in the handling of the brass in *Lemminkäinen in Tuonela*. The sense of national identity comes from Sibelius's interest in folksong. He visited Karelia in 1892 and, though he took little scholarly interest in collecting folk music, he heard the runic singing of Larin Paraske during the early

1890s. Its impact on him was considerable: its rhythmic structure and modal habits of mind coloured his own melodic thinking and became part of him; there was no attempt to absorb folk melodies consciously or subject them to harmonization in the course of symphonic or other major works.

After the cool reception accorded the *Lemminkäinen* suite, Sibelius suffered another setback. Faltin's attempt to secure him the post of professor of composition at the music school on his own retirement misfired. Although the recommendation was endorsed, an appeal to St Petersburg overturned the decision in favour of Kajanus; hence the state pension which the Finnish senate voted Sibelius some months later in 1897 was partly an acknowledgment of his creative achievements to date, but also a gesture to meet the body of opinion which felt Sibelius had been badly treated. Some years later this was turned into a life pension, but Sibelius's finances remained precarious until the 1920s; the sum involved was a small one and Sibelius's tastes and appetites were expensive. His debts reached 100,000 marks at one point during the first decade of the century, though his friend Baron Axel Carpelan, not himself a wealthy man, arranged a certain amount of anonymous patronage. It was not until the institution of performing rights in the 1920s that Sibelius's finances were secured.

CHAPTER THREE

International renown

In 1898 Sibelius started working on his first non-prog-
rammatic symphony. Quite possibly the successful performance of a symphony by Ernst Mielck, which
Kajanus presented the previous year, acted as a spur:
Mielck was younger, and his was the first Finnish symphony since Ingelius's. No doubt Flodin's response to
the *Lemminkäinen* suite reinforced Sibelius's recognition of the strength of his purely symphonic instinct.
The success of its first performance in 1899 was sur-
passed only by that of the music he composed later the
same year for a pageant mounted in connection with the
press pension celebrations. Though the pretext was in-
nocuous enough, the pageant became a rallying point for
patriotic sentiment at a time when the tsarist hold on
Finland was tightening. One of its numbers became the
well-known *Finlandia*.

If the late 1890s had seen the consolidation of
Sibelius's position as Finland's leading composer, the
next decade was to see the growth of his international
reputation. In 1898, through the good offices of his
friend Adolf Paul, a pianist *manqué* and the author
of the play *Kung Kristian II* to which Sibelius had
composed incidental music, he had been able to acquire a
continental publisher, Breitkopf & Härtel. To them
Sibelius was later to sell the rights of the *Valse triste*
for the derisory sum of 300 marks. In 1900 Kajanus

took the Helsinki PO, which he had founded only 18 years earlier, on its first European tour, which culminated in their appearance at the Paris World Exhibition. *The Swan of Tuonela* and *Lemminkäinen's Return*, as well as *Finlandia* and the First Symphony, were well received; in 1901 Sibelius was invited to Heidelberg to conduct his music, and the following year he was invited to Berlin to conduct the definitive version of *En saga* at one of Busoni's new music concerts. His fame was not confined to Germany: Henry Wood conducted Sibelius's *Kung Kristian II* suite at a promenade concert in 1901, and Granville Bantock introduced the First Symphony to English audiences four years later. Sibelius himself travelled extensively on the Continent during these years, meeting Dvořák when he passed through Prague in 1901, and spending the beginning of the following year in Rapallo, where the first ideas for the Second Symphony came to him. His sketchbooks rapidly filled out with various projects including a work on the Don Juan theme called 'Festival: Four Tone-poems for Orchestra' and he toyed with the idea of writing a work based on the *Divine Comedy*. In his sketches the main theme of the slow movement of the symphony bears the superscription 'Don Juan' and the second idea, in F♯ major, the word 'Christus'. None of the symphonies has enjoyed such immediate success as the Second. All four performances in 1902 were completely sold out and the symphony began its triumphant pilgrimage round the world. Some of its success was extra-musical in that the work struck a responsive patriotic nerve at a time when Finnish autonomy was threatened with erosion and nationalist fervour was at its height. If it inhabits much the same world as the First, it views it through more subtle and

refined lenses. As in the First Symphony, it is the opening movement that makes the most profound impression. Its very air of relaxation and effortlessness serves
to mask its inner strength.

The period of the Violin Concerto (1903, rev. 1905)
was one of some turbulence, with mounting debts and
bouts of heavy drinking which presented considerable
domestic strain. In 1904 Sibelius bought a plot of land
at Järvenpää, not far from Helsinki, where a villa was
built in which he spent the rest of his life. The next few
years were enormously productive and important: they
found him turning away from the national romanticism
of his first period, and at the same time turning his back
on the increasing complexity and richness of many of
his contemporaries during the first decade of the century. Indeed, the Third Symphony, begun in 1904 but
not completed until 1907, evinces a remarkable neo-
classicism in both content and form. Gerald Abraham
has said of its opening movement: 'In clearness and
simplicity of outline, it is comparable with a Haydn or
Mozart first movement ... nevertheless, the organic
unity of the movement is far in advance of anything in
the classical masters; and even the general architecture
is held together in a way that had classical precedents
but had never before been so fully developed'. In the
Third Symphony and its remarkable successor Sibelius
refined and distilled his musical language so that the
rhetoric of his early romantic style was replaced by an
economy of means and purity of utterance far removed
from the opulent orchestral colours of Strauss, Ravel
and Skryabin, or the exploratory world of sound and
extension of expressive devices that mark composers
like Schoenberg and Bartók.

The increased concentration of the outer movements of

the symphony can also be seen in *Pohjolan tytär* ('Pohjola's daughter'), in which symphonic and programmatic elements are fused in perfect balance. Some of Sibelius's finest incidental music, to Maeterlinck's *Pelléas et Mélisande* and Hjalmar Procopé's *Belsazars gästabud* ('Belshazzar's feast'), dates from this period.

A serious illness developed in 1908, and Sibelius underwent a series of operations in Helsinki and Berlin for suspected cancer of the throat. For a number of years he was forced to abjure alcohol and cigars, and the bleak possibilities which the illness opened up may well have served to contribute to the austerity, concentration and depth of the works which followed in its wake, the Fourth Symphony, *Barden* ('The bard') and *Luonnotar*. In the Fourth Symphony (1911), his musical language is more elliptic, and in his extensive use of the tritone he came closer to indeterminate tonality than in any other major work. He continued to travel extensively during the period up to the outbreak of World War I, making several visits to England (1905, 1908, 1909 and 1912); he continued to receive marks of international recognition, including the Légion d'honneur (1906), the offer of the chair of composition at the Imperial Academy of Music, Vienna (1912), and an honorary doctorate from Yale (1914). For his American visit in 1914 he composed *Aallottaret* ('The oceanides'), which he conducted at Norfolk, Connecticut. He derived great satisfaction from these appearances as an executant, which perhaps compensated for the failure of his ambitions as a violin virtuoso. He encountered directly the extent of his popularity in the USA, but, though he toyed with the idea of returning to make a concert tour to pay off his debts, and though he was offered the post of director of the Eastman School after the war, he never revisited America.

The war period saw no major compositions apart from the Fifth Symphony. One of the consequences of the hostilities was the loss of revenue from Breitkopf & Härtel, and as a result (and in the absence of conducting commitments abroad) Sibelius composed large quantities of piano music and of violin and piano miniatures for the domestic market. The Fifth Symphony gave Sibelius more trouble than any other work. He mentioned it for the first time as early as 1912 in his diary and in a curiously tentative first version it was given on his 50th birthday on 8 December 1915, an event treated almost as a national holiday. In its original form, the orchestral parts of which survive, the work was in four movements: Sibelius linked the first two for a performance in 1916, though only a double bass part of this version has come to light. An interesting recent discovery, however, is a sketchbook, begun some time in 1914 (after Sibelius's return from America) and covering the period up to mid-June 1915, from which one can follow the way the basic ideas of the Fifth Symphony gradually formed. It is well known that the Sixth and Seventh Symphonies were conceived at the same time and have distinct thematic affinities, but it now transpires that some of the ideas for the Fifth Symphony found an eventual outlet in the Sixth. Indeed the sketches for the opening of the Sixth in 1915 do not really undergo any significant change; the essentials are easily discernible and the key is clearly D minor. (The sketchbook contains ideas for works other than the symphonies; Sibelius was planning a second violin concerto at this time.) The famous peroration leading into the coda of the finale of the Fifth was also sketched in D minor. Sibelius had intended to make a further revision in 1917 for his brother-in-law, Armas Järnefelt to conduct in Stockholm, but the deteriorating situation in Finland preoccupied him and the work was not put into its definitive form until 1919.

After the October Revolution in Russia Finland proclaimed its independence, but early in 1918 the Red Guards attempted a *coup d'état* and Finland was plunged into civil war. Sibelius fled from Järvenpää, as his sympathies were with the Whites, and moved to the Lapinlahti hospital, where his brother was a senior doctor.

CHAPTER FOUR

Last works

It was between the years 1920 and 1925 that Sibelius's active creative life came virtually to a close. Apart from a considerable quantity of light music (he longed to repeat the success of *Valse triste* for his own rather than his publisher's advantage), some of it of good quality, the postwar years saw only four major works: the Sixth and Seventh symphonies, the incidental music to *The Tempest*, and his final utterance, *Tapiola*. It coincided with the end of his public career as a performer: he visited London for the last time in 1921 to conduct the Fourth and Fifth symphonies, and was reunited with Busoni; he conducted his Second Symphony in Rome, his Sixth in Göteborg and the Seventh in Stockholm. He survived his last important works by more than 30 years, living in retirement in his villa, Ainola, in Järvenpää.

The reasons for the silence of the last years are numerous. Artistically Sibelius felt isolated from the direction which music was taking: he had little sympathy with Les Six, Schoenberg or Stravinsky. Among the avant garde of the day, Bartók strongly appealed to him, but in other respects his sympathies were for composers less exploratory in language and whose musical resources drew on traditional symphonic thinking. Moreover, given the achievement of *Tapiola*, its uncompromising vision, its desolate, unpeopled world, it is

26. Sibelius at his villa, Ainola, in Järvenpää

difficult to envisage the course his music could have taken. At a personal level he was beginning to sense isolation: among his friends he suffered a grievous loss in the death of his artistic confidant Axel Carpelan. And for all his growing success in the Anglo-Saxon world, the failure of his music, which had enjoyed powerful and persuasive advocacy in Wilhelmine Germany, to regain its pre-war position on the Continent was a source of undoubted concern. In the late 1920s he worked on an Eighth Symphony which, according to a letter to Olin Downes of 19 January 1931, was nearing completion that year. Part of the score was delivered for copying in 1933 (a note from the copyist has come to light with a draft reply from Sibelius from which it seems that the symphony would have been about the same size as the Second) but was subsequently retrieved. Three bars of a sketch marked 'Sinfonia 8 Commincio' survive but reveal as little as would the first three bars of the Second Symphony, were only that to remain of the finished score. The composer Joonas Kokkonen advanced the theory that *Surusoitto* ('Funeral'), the second of the two pieces for organ op.111 (composed on the death of the painter Akseli Gallén-Kallela, his friend and drinking-companion of the 1890s), might in fact be drawn from the symphony. During the height of the Finnish civil war Sibelius resumed drinking. In fact by the late 1920s and early 1930s, he had developed quite a tremor and writing was burdensome. Kokkonen argued that it would be more natural for him to 'raid' the score of the Eighth Symphony and perhaps adapt a few paragraphs from it, a view to which his widow lent support.

The vogue for Sibelius in the 1930s and claims made by Cecil Gray and Constant Lambert on his behalf, combined with his innate propensity for self-doubt, served to hold up publication. Eventually the work disappeared or was destroyed. Throughout the 1930s Sibelius never left Finland and thus did not encounter at first hand the fame and veneration he enjoyed in England and America. The Sibelius Society was founded; Kajanus recorded some of his music; and complete cycles of the symphonies were conducted by Koussevitzky and Beecham. Apart from a brief period under Hitler, when his music was mobilized by the Nazis for ideological purposes, thus fostering subsequent hostility, Germany continued its indifference, in spite of the advocacy of Rosbaud and Karajan, and aided by the antipathy of Adorno and Thomas Mann. Elsewhere, however, Sibelius continued to bask in great popular esteem: on his 90th birthday he received 1200 telegrams, presents from all the Scandinavian monarchs, cigars from Churchill, tapes from Toscanini and the inevitable plethora of concerts and other festivities. He died at his home in Järvenpää from a cerebral haemorrhage on 20 September 1957, less than three months short of his 92nd birthday.

Even before his death the inevitable reaction against the dominant position Sibelius occupied in the Anglo-Saxon world had begun to manifest itself. There was a slackening of serious critical interest: only one study of his music appeared in the 1950s after Gerald Abraham's symposium, and he began to encounter journalistic hostility on both sides of the Atlantic. The enthusiasm of Downes, Gray, Lambert and others sharpened the intensity of the normal reaction, and the tendency, as evinced in Johnson's book, was 'to cut the composer down to size'. However, Sibelius was one of the first composers in whose popularity the gramophone had played a decisive role, and the volume of recordings continued unabated. The complete cycles of the symphonies by younger conductors such as Bernstein and Maazel during the 1960s, as well as the rediscovery of such works as *Kullervo*, which Sibelius had banned in his lifetime, were in the long run no less important than the work of Koussevitzky and Beecham in maintaining interest in the composer and promoting the further dissemination of his art.

CHAPTER FIVE

Orchestral achievement

Sibelius's achievement rests mainly on his symphonies and symphonic poems. In both he showed an exceptional sense of form and mastery of the orchestra. Each of the symphonies shows a fresh approach to the symphonic challenge; there is no archetypal Sibelius symphony from which one can distil a set of rules or a formula that applies to all. Moreover, Sibelius's development from one work to the next was less predictable than that of almost any other 19th- or early 20th-century symphonist: from the vantage point of one it would have been impossible to foresee the character and structure of the next. In terms of their spiritual progress, the journey from no.1 to no.7 is greater than that between Brahms's no.1 and no.4, even if such predecessors as Brahms, Tchaikovsky and Dvořák may have encompassed a wider range of human experience and possessed other dimensions denied Sibelius. At the same time the voyage from the climate of Slav romanticism that fostered the First Symphony into the wholly isolated and profoundly original world of the Sixth and Seventh, at a time when the general mainstream of music was moving in other directions, was one of courageous spiritual discovery. *Tapiola*, in which Sibelius's work culminated, united this symphonic progress with the lifelong preoccupation he had with nature and myth. In the 1930s Gray made sweeping claims for it: 'Even if Sibelius had written nothing else, this one work would entitle him to a place among the greatest masters of all time'. Certainly Sibelius exhibited in *Tapiola* a most subtle command of symphonic procedure and achieved a continuity of thought paralleled only in the symphonies. It is unique even in Sibelius's own output; its world is new and unexplored, a world of strange new sounds, a landscape that no other tone poem has painted with such inner conviction and identification. It is to the northern forests what *La mer* is to the sea.

Its predecessor, the Seventh Symphony, illustrates the truth of the assertion that Sibelius never approached the symphonic problem twice in the same way. If each of the symphonies shows a continuing search for new formal means, it is not because of any conscious commitment to structural innovation, but rather the result of a capacity to allow each idea to dictate the flow of the music and establish its own disciplinary logic. The Seventh consummates the 19th-century search for symphonic unity: it is in one continuous movement, and though there are passages with the character of a scherzo or a slow section, it is impossible to define exactly where one section ends and another begins, so complete is Sibelius's mastery of transition and control of simultaneous tempos. The Seventh Symphony came as the climax of a lifetime's work: it has the effect of a constantly growing entity, in which the thematic metamorphosis works at such a level of sophistication that a listener is barely aware of it. In its degree of organic integration and thematic working, the piece stands at the peak of the symphonic tradition. It sums up the Sibelian approach to the symphony as outlined in Mahler's oft-quoted remark, made when he visited

27. *Autograph MS from Sibelius's Symphony no.7 in C, composed 1924*

Helsinki in 1907: 'I said that I admired its style and severity of form, and the profound logic that created an inner connection between all the motifs'.

The Sixth Symphony (1923) has excited the admiration of Sibelius connoisseurs but remains the least performed of the seven. Ralph Wood called it Sibelius's greatest symphony, 'a dazzling display of a technique so personal and so assured that its very achievements are hidden in its mastery and in its entire synthesis with its subject matter'. The work inhabits a world as far removed from the Fifth as that in its turn is from the Fourth. If it lacks the heroic countenance of the Fifth and the stern epic majesty of the Seventh, it possesses a purity of utterance and spirit that has few parallels in either Sibelius or the music of its time. Although the first movement of the Fifth (final version 1919) was originally in two separate entities, the organic relationship of their material was established and Sibelius fashioned the work into one continuous movement. Simon Vestdijk spoke of it as his finest individual symphonic movement, and a comparative study of the two versions affords a valuable insight into the way in which the organic elements that bind the two movements were slow to reveal their full implications to the composer. The finale is less concentrated and its scoring less effective.

Although Sibelius was late in coming to the orchestra, his mastery of its intricacies came relatively rapidly. In this he was helped by his friendship with Kajanus and by his own conducting experience. By the end of the 1890s he was able to establish within a few seconds a sound world entirely his own. As with Berlioz, his thematic inspiration and its harmonic clothing were conceived directly in terms of orchestral sound; the substance and the sonority were indivisible. Many of the characteristics of Sibelius's orchestral style are instantly recognizable: the 'cross-hatch' string writing, woodwind instruments in 3rds, long sustained brass chords which open *sforzato subito pianissimo* and then make a slow crescendo, the long pedal points and the openness of the textures, all of which show a thorough assimilation of Wagner and Berlioz, and the transmutation of such influences into a wholly individual expressive technique. Sibelius's orchestral mastery is as much in evidence in his powerful works for the theatre as in the symphonies and symphonic poems. His theatre music, which is much more loosely structured than the rest of his orchestral output, depends on the intensity with which he could distil an atmosphere, and this in turn depended in no small measure on his genius for realizing the potentialities of the orchestra. The inner movements of *Belshazzar's Feast* are not densely composed, but a magical atmosphere is created by the simplest musical and orchestral devices. Much the same can be said, at a higher level, of his music for the Copenhagen production of *The Tempest*.

His early years as a violinist surely acquainted him with the full range of expressive devices a string instrument can command; and, whether he wrote for the strings in an orchestral context in, say, *The Tempest* or *The Oceanides*, or for strings alone, as in *Rakastava* ('The lover'), the result is unfailingly imaginative and resourceful. Writing for the solo violin, in the concerto and the delightful *Humoresques* opp.87 and 89, he was more conditioned by the school of virtuoso violinists in which he was brought up. These works do have moments of depth and, more particularly, so do the Two Serenades op.69, but, as with the piano music, the music he provided for domestic use was less characteristic and distinguished. More puzzling, considering the early (and exclusive) preoccupation with chamber music, is his evident lack of interest in the string quartet. Only one work, albeit a fine one, the String Quartet 'Voces intimae', comes from his maturity, though he toyed with plans for others. Indeed, parts of the Fourth Symphony were originally sketched for string quartet.

CHAPTER SIX

Other works

Sibelius composed more than a hundred songs, and the neglect of the finest of them must be because of the comparative inaccessibility of the Swedish and Finnish languages to the non-native singer, and to the extent to which the orchestral music overshadows the rest of his output in the public mind. The finest songs, *Jubal*, *Theodora*, *På verandan vid havet* ('On a balcony by the sea') and the bulk of the opp.36–8 collections, written at the turn of the century, exhibit a striking ability to evoke atmosphere by the simplest means, comparable to that shown in the best of the incidental scores. Sibelius's activity as a song composer extended over most of his creative life. Though he chose Finnish texts for his choral works and many of his partsongs, most of his solo songs sprang from his love for the Swedish nature poets, and in particular Runeberg, settings of whom dominate his early output (opp.3, 13 and 17). His contact with such singers as Ida Ekman and Aïno Ackté, for whom *Luonnotar* was composed, further deepened his understanding of vocal writing. Ekman sang op.17, which includes such widely contrasting songs as *Vilse* ('Astray') and the powerful and forward-looking *Se'n har jag ej frågat mera* ('Since then I have questioned no further'), to Brahms, and both she and Ackté championed the songs on the Continent during the period 1890–1910.

The songs in general range from the smallest miniature, strophic and diatonic in character (like *Astray* and *Jägargossen*), to the large-scale nature songs like *Autumn Evening*, which evoke a world more closely akin to the orchestral music. Indeed, *Autumn Evening* is more deeply characteristic of the austere, uncompromising nature poet of *The Bard* and the Fourth Symphony than almost any other of his songs; Rydberg's poem dealing with the solitary wanderer exulting in the power and majesty of nature struck a particularly responsive note in the composer. For Sibelius, the relationship with nature seems to have been predominant: nature and myth interested him more deeply than the tensions arising from human relationships. His songs have none of the psychological dimensions of Wolf or Musorgsky, and he was undoubtedly inhibited as a song composer by his somewhat limited command of keyboard idiom. Rarely does the piano comment on the vocal line with the freedom of the great lieder composers or even of his countryman Kilpinen. Yet *Autumn Evening* and its companion Rydberg settings in op.38, as well as his extraordinary masterpiece *Luonnotar*, are evidence of an original approach to the voice, and even some of his less inspired songs still show a highly developed feeling for the inflections and spirit of the Swedish poets he set.

Some of the best known, such as *Svarta rosor* ('Black roses') and *Flickan kom ifrån sin älsklings möte* ('The maiden's tryst') are closely related to the salon taste of the 1890s, and their popularity has done the remainder no service. Sibelius's most inspired vocal writing comes from 1909–10; the op.35 songs and *Luonnotar* have a boldness and freedom of line and an

300

28. Jean Sibelius, 1955

imaginative sweep that raise them to the level of the finest of his orchestral works. *Luonnotar*, in particular, employs a wide range (the work was written for Ackté) and falls into a special category. Half song, half symphonic poem, it was begun after op.35, probably as early as 1910, when Sibelius's thoughts were turning to the Fourth Symphony. Its wide leaps make cruel demands on the soloist and, though at first sight the line appears almost instrumental in conception, it does in fact display an uncanny feeling for the language (this is one of Sibelius's rare large-scale solo vocal works in Finnish). In conception and design it is symphonic: its power and intensity are paralleled only in *Tapiola* and *The Bard*. After the fine Tavaststjerna setting *Långsamt som kvällsyn* ('Slowly as the evening sun'), the first of the op.61 set, there are few Sibelius songs of outstanding quality. The opp.86 and 88 sets are relatively routine and, though there are isolated exceptions (the 1918 setting of Gripenberg's *Narciss*, for example), the later songs are weaker.

Sibelius had relatively little feeling for the piano, but he composed prolifically for it throughout his creative career. Rarely is his keyboard writing idiomatic, though his output contains a handful of fine miniatures, including the Three Sonatinas op.67, probably his best piano work. Generally speaking his keyboard layout is ineffective; he employed a limited range of devices and made little attempt to exploit the sonorities and colouristic effects of the extremities of the keyboard. Much of his output was designed for the home rather than the concert hall and was the outcome more of financial considerations than of a strong creative impulse. The D♭ Romance of op.24 and other salon pieces served these purposes and have not maintained a place in the repertory. Sibelius's piano writing in general either reflected orchestral habits of mind transplanted to the keyboard or had a pallor more characteristic of 19th-century Scandinavian nationalism than of his own personality. Unlike Nielsen he never developed a personal keyboard style, and only in the sonatinas is there a strong hint of the concentration and economy that mark his finest orchestral scores.

The immediate influence of Sibelius was felt in the Scandinavian countries, on his countryman Madetoja, on Swedish composers as diverse as Stenhammar, Rosenberg and Wirén, and also in the Anglo-Saxon countries, particularly during the 1930s: Bax, Moeran, the Symphony in B♭ minor of Walton, and the First Symphony of Barber all showed a debt to him. At a deeper level, that of structure rather than idiom, Sibelius's music has influenced the symphonies of Harris and Holmboe. Sibelius is the only commanding figure that Finland has produced. His close identification with the cause of Finnish nationalism and his preoccupation with its mythology and nature give him a special place in the history of Scandinavian music, for he enshrines the spirit of the north in a unique fashion. Were this his sole achievement he would only be a colourful nationalist like Grieg or Delius. It is his highly developed feeling for form and extraordinary originality as a symphonic thinker that assures him of a secure position in the history of European music. His ultimate conviction that 'classicism is the way of the future' may not seem to have much supporting evidence, but future generations of composers may well look to his example of independence.

WORKS

Numbers in the right-hand column denote references in the text.

STAGE

op.		pages
—	Näcken [The watersprite] (2 songs for fairy play, G. Wennerberg), 1888, unpubd	297
—	Karelia (incidental music), 1893, unpubd; ov., op.10, 1893; suite; op.11, 1893	277
—	Jungfrun i tornet [The maiden in the tower] (opera, 1, Hertzberg), 1896, unpubd	277
8	Ödlan [The lizard] (incidental music, M. Lybeck), 1909, unpubd	
27	Kung Kristian II (incidental music, A. Paul), 1898; suite, 1898	281, 282
—	Press Pension Celebrations (music for pageant), 1899	281
44	Kuolema [Death] (incidental music, A. Järnefelt), 1903, unpubd; 1 no. rev. as Valse triste, 1904; 2 nos. rev. as Scene with Cranes, 1906, unpubd; also 2 addl pieces, op.62, for new production, 1911	281, 287
46	Pelléas et Mélisande (incidental music, Maeterlinck), 1905; suite, 1905	284
51	Belshazzars gästabud [Belshazzar's feast] (incidental music, H. Procopé), 1906, unpubd; suite (1907)	284, 297
54	Svanevit [Swanwhite] (incidental music, Strindberg), 1908, unpubd; suite (1909)	
60	Twelfth Night (2 songs, Shakespeare), 1v, gui/pf, 1909: Kom nu hit, död [Come away, death], also arr. 1v, harp, str orch, unpubd; Och när som jag ver en liten smådräng [When that I was and a little tiny boy]	
—	Die Sprache der Vögel (wedding march for play, Paul), 1911, unpubd	
71	Scaramouche (music for tragic pantomime, P. Knudsen, M. T. Bloch), 1913	
83	Jokamies [Everyman] (incidental music, Hofmannsthal), 1916, unpubd	
109	The Tempest (incidental music, Shakespeare), 1925; prelude, 2 suites	287, 297

ORCHESTRAL

op.		pages
—	Overture, E, 1890–91, unpubd	
—	Ballettikohtaus [Ballet scene], 1891, unpubd	292–8
6	Cassazione, 1895, rev. 1904, unpubd	
9	En saga, 1892, rev. 1902	275, 277, 278, 279, 282
10, 11	Karelia: see 'Stage'	277
—	Menuetto, 1894, unpubd	
15	Skogsrået [The wood nymph], tone poem, 1895, unpubd	
16	Vårsång [Spring song], tone poem, 1894	
22	Lemminkäis-sarja [Lemminkäinen suite]: Lemminkäinen ja saaren neidot [Lemminkäinen and the maidens of the island], 1895, rev. 1939; Lemminkäinen Tuonelassa [Lemminkäinen in Tuonela], 1895, rev. 1897, 1939; Tuonelan joutsen [The swan of Tuonela], 1893, rev. 1897, 1900; Lemminkäisen paluu [Lemminkäinen's return], 1895, rev. 1897, 1900	277, 278, 279, 280, 281, 282
—	Tiera, tone poem, brass, perc, 1898	
25	Scènes historiques I [from Press Pension Celebrations music], 1899, rev. 1911	
26	Finlandia [from Press Pension Celebrations music], 1899, rev. 1900	281, 282
39	Symphony no.1, e, 1899	277, 279, 281, 282, 283, 291, 292
42	Björneborgarnes March, 1900, unpubd	
43	Romance, C, str, 1903	282–3, 287, 289, 291
43	Symphony no.2, D, 1901–2	
—	Overture, a, 1902, unpubd	
45/1	Dryadi [The dryad], tone poem, 1910	
47	Violin Concerto, d, 1903, rev. 1905	283, 297
49	Pohjolan tytär [Pohjola's daughter], sym. fantasia, 1906	284
52	Symphony no.3, C, 1907	
53	Pan and Echo, dance intermezzo, 1906	283–4, 291
55	Öinen ratsastus ja auringonnousu [Nightride and sunrise], tone poem, 1908	
59	In memoriam, funeral march, 1909	
—	Balettikohtaus [Ballet scene], 1909, unpubd	
63	Symphony no.4, a, 1911	283, 284, 287, 291, 296, 298, 300, 301
64	Barden [The bard], tone poem, 1913, rev. 1914	284, 300, 302

66 Scènes historiques II, suite, 1912 — 298
69 Two Serenades, vn, orch, D, g, 1912–13 — 284, 297
73 Aallottaret [The oceanides], tone poem, 1914 — 285, 286, 287, 291, 298
77 Two Pieces, vn/vc, orch, 1914 — 297
82 Symphony no.5, E♭, 1915, rev. 1916, 1919 — 297

87 Humoresques nos.1–2, d, D, vn, orch, 1917
89 Humoresques nos.3–6, g, g, E♭, g, vn, orch, 1917
— Promotiomarssi [Academic march], 1919, unpubd
96/1 Valse lyrique, 1920
96/2 Autrefois (Scène pastorale), orch, 2 solo vv ad lib, 1919
96/3 Valse chevaleresque, 1920
98/1 Suite mignonne, fl, str, 1921
98/2 Suite champêtre, str, 1921
100 Suite caractéristique, harp, str, 1922
104 Symphony no.6, d, 1923 — 285, 287, 291, 292, 296

105 Symphony no.7, C, 1924 — 285, 287, 291, 292, 293, 294–5, 296

112 Tapiola, tone poem, 1926 — 287, 292–3, 302
— Symphony no.8, 1929–34, destroyed — 289–90

See also suites from stage works and arrs. of other compositions

CHORAL

7 Kullervo (Kalevala), sym., S, Bar, male vv, orch, 1892 — 275, 276, 277, 279, 291
14 Rakastava [The lover] (Kanteletar), male vv, 3 nos., 1893; arr. male vv, str orch, 1894, unpubd; arr. mixed vv, 1898; recomposed for str orch, triangle, timp, 1911 — 297
— Cantata for the Helsinki University ceremonies of 1894 (K. Leino), mixed vv, orch, 1894, unpubd
18 Nine Partsongs (Aho, Kanteletar, Kivi, Kalevala), TTBB, 1893–1904
19 Impromptu (Rydberg), female vv, orch, 1902, rev. 1910
21 Natus in curas (Gustafsoon), TTBB, 1896
— Cantata for the Coronation of Nicholas II (Cajander), solo vv, mixed vv, orch, 1896, unpubd
— Aamusumussa [Morning mist] (Erkko), 3 children's vv, 1896
— Työkansan marssi [Workers' march] (Erkko), mixed vv, 1893–6
23 Cantata for the Helsinki University ceremonies of 1897 (Koskimies), solo vv, mixed vv, orch, 1897, unpubd; 10 songs excerpted (1899)

28 Sandels (Runeberg), improvisation, male vv, orch, 1898, rev. 1915
29 Snöfrid (Rydberg), improvisation, reciter, male vv, orch, ?1900
30 Islossningen i Uleå älv [The breaking of the ice on the River Uleå] (Z. Topelius), improvisation, reciter, male vv, orch, 1898
31/1 Laulu Lemminkäiselle [A song for Lemminkäinen] (Veijola), mixed vv, orch, ?1894, unpubd
31/2 Har du mod? (Wecksell), male vv, orch, 1904
31/3 Atenarnes sång [Song of the Athenians] (Rydberg), boys' vv, male vv, wind, perc, 1899; arr. with orch, unpubd
— Carminalia (Lat. student songs), SAB/S, A, pf, harmonium, 1899
— Kotikaipaus [Nostalgia] (von Konow), 3 female vv, 1902
— Til Thérèse Hahl (Wasatjerna), mixed vv, 1902
32 Tulen synty [The origin of fire] (Kalevala), Bar, male vv, orch, 1902, rev. 1910
— Ej med klagan [Not with lamentations] (Runeberg), mixed vv, 1905
48 Vapautettu kuningatar [The liberated queen] (Cajander), cantata, mixed vv, orch, 1906
— Kansakoululaisten laulu [Folk school children's march], children's vv, 1910
— Cantata (von Konow), female vv, 1911
65 Two Partsongs (E. W. Knape, Engström), mixed vv, 1911–12
— Drömmarna [Dreams] (Reuter), mixed vv, 1912
— Uusmaalaisten laulu [Song for the people of Uusimaa] (Terhi), mixed vv, 1912
— Three Songs for American Schools (Dixon, Scott, Macleod), 1913
84 Five Partsongs (Fröding, Gripenberg, Reuter), male vv, 1914–15
— Kuutamolla [In the moonlight] (Suonio), male vv, 1916
91/1 March of the Finnish Jaeger Battalion (Nurmio), male vv, 1917; arr. male vv, orch, ?1917
91/2 Scout March (Procopé), mixed vv, 1917; arr. mixed vv, orch, ?1917
— Fridolins dårskap [Fridolin's folly] (Karlfeldt), male vv, 1917
— Brusande rusar en våg [The roaring of a wave] (Schybergson), male vv, 1918
— Jone havsfärd [Jonah's voyage] (Kaarlfeldt), male vv, 1918
— Ute hörs stormen [One hears the storm outside] (Schybergson), male vv, 1918
92 Oma maa [Our native land] (Kallio), cantata, mixed vv, orch, 1918
93 Jordens sång [Song of the earth] (Hemmer), cantata, mixed vv, orch, 1919

Jag ville jag vore i Indialand [I wish I dwelt in India land] (Fröding), 1904

— Erloschen [Extinct] (Busse-Palmo), 1906

50 Six Songs, 1906:
  Lenzgesang (A. Fitger)
  Sehnsucht (R. Weiss)
  Im Feld ein Mädchen singt (M. Susman)
  Aus banger Brust (Dehmel), orchd
  Die stille Stadt (Dehmel)
  Rosenlied (A. Ritter)

57 Eight Songs (Josephson), 1909:
  Älvan och snigeln [Älvan and the snail]
  En blomma stod vid vägen [A little flower in the path]
  Kvarnhjulet [The millwheel]
  Maj [May]
  Jag är ett träd [The bare tree]
  Hertig Magnus [Duke Magnus]
  Vänskapens blomma [The flower of friendship]
  Necken [The elf king]

61 Eight Songs, 1910:
  Långsamt som kvällsyn [Slowly as the evening sun] (Tavaststjerna) 302
  Vattenplask [Lapping waters] (Rydberg)
  När jag drömmer [When I dream] (Tavaststjerna)
  Romeo (Tavaststjerna)
  Romance (Tavaststjerna)
  Dolce far niente (Tavaststjerna)
  Fåfäng önskan [Idle wish] (Runeberg)
  Vårtagen [Spell of springtime] (Gripenberg)

72 Six Songs:
  Vi ses igen [Farewell] (Rydberg), 1914
  Orions bälte [Orion's belt] (Topelius), 1914
  Kyssen [The kiss] (Rydberg), 1915
  Kaiutar [The echo nymph] (Kyösti), 1915
  Der Wanderer und der Bach (M. Greif), 1915
  Hundra vägar [A hundred ways] (Runeberg), 1907

86 Six Songs, 1916:
  Vårförnimmelser [The coming of spring] (Tavaststjerna) 302
  Längtan heter min arfvedel [Longing is my heritage] (Karlfeldt)
  Dold förening [Hidden union] (C. Snoilsky)
  Och finns det en tanke? [And is there a thought?] (Tavaststjerna)
  Sångarlön [The singer's reward] (Snoilsky)

  I systrar, i bröder [Ye sisters, ye brothers] (Lybeck) 302

88 Six Songs, 1917:
  Bläsippan [The anemone] (F. M. Franzén)
  De bägge rosorna [The two roses] (Franzén)
  Hvitsippan [The star-flower] (Franzén)
  Sippan [The primrose] (Runeberg)
  Törnet [The thorn] (Runeberg)
  Blommans öde [The flower's destiny] (Runeberg)

90 Six Songs (Runeberg), 1917:
  Norden [The north]
  Hennes budskap [Her message]
  Morgonen [The morning]
  Fågelfångaren [The bird catcher]
  Sommarnatten [Summer night]
  Hvem styrde hit din väg? [Who has brought you here?]

— Narciss (Gripenberg), ?1918 302
— Små flickorna [Small girls] (Procopé), 1920

OTHER VOCAL WORKS
(vocal orchestral)

— Serenade (Stagnelius), Bar, orch, 1895, unpubd
33 Koskenlaskijan morsiamet [The rapids-shooter's brides] (A. Oksanen), Bar/Mez, orch, 1897
70 Luonnotar (Kalevala), tone poem, S, orch, ?1910 284, 299, 300, 301–2

(duet)
— Tanken [The thought] (Runeberg), 2 S, pf, 1915, unpubd

(music for recitation)
— Tränaden [Longing] (Stagnelius), pf, 1887, unpubd
— Svartsjukans nätter [Nights of jealousy] (Runeberg), pf, trio, 1888, unpubd
15 Skogsrået [The wood nymph] (Rydberg), pf, 2 hn, str, 1894, unpubd; rev. as tone poem
— Grefvinnans konterfej [The countess's portrait] (Topelius), str orch, 1906, unpubd
— Ett ensamt skidspår [The lonely ski trail] (Gripenberg), pf, 1925; arr. harp, str orch, 1948, unpubd

CHAMBER AND INSTRUMENTAL
Unpubd early works: Vattendroppar [Water drops], vn, vc, 1876; Pf Trio, a, 1881–2, frags.; Pf Qt, e, 1881–2; Vn Sonata, d, 1881–2; Andantino, vc, pf, ?1884; Str Qt, Eb, 1885; Str Trio, g, ?1885; Pf Trio 'Korpo', 1887; Andante cantabile, vn, pf,

285, 298
271
272
279

	1887; Qt, g, harmonium, pf trio, c1887; Pf Trio 'Lovisa', C, 1888; Theme and Variations, c♯, str qt, 1888; Pf Qnt, g, 1889; Suite, A, str trio, 1889; Vn Sonata, F, 1889; Andantino, Menuetto, cl, 2 cornets, 2 hn, bar, tuba, 1890–91; Qt, C, 2 vn, vc, pf, 1891; Rondo, va, pf, 1893	275 279
2	Two Pieces, vn, pf, 1888, rev. 1912	
4	String Quartet, B♭, 1889–90, unpubd; Presto arr. str orch, unpubd	275, 279
—	Kehtolaulu [Lullaby], vn, kantele, 1899	
—	Fantasia, vc, pf, 1900	
20	Malinconia, vc, pf, 1901	
56	String Quartet 'Voces intimae', d, 1909	298
78	Four Pieces, vn/vc, pf, 1915–19	
79	Six Pieces, vn, pf, 1915	
80	Sonatina, E, vn, pf, 1915	
81	Five Pieces, vn, pf, c1915	
—	Andante festivo, str qt, 1922; arr. str orch, timp ad lib	
102	Novelette, vn, pf, 1923	
106	Cinq danses champêtres, vn, pf, 1925	289
111	Two Pieces, org, 1925, 1931	
115	Four Pieces, vn, pf, 1929	
116	Three Pieces, vn, pf, 1929	

PIANO

—	Unpubd early works: Au crépuscule, 1887; Andantino, 1888; Scherzo, ?1888; Allegretto, 1889; Florestan, suite, 1889	285, 302
5	Six Impromptus, 1893; 2 nos. arr. str orch	
12	Sonata, F, 1893	

—	Kavaljeren [The cavalier], 1900	
—	[6] Finnish Folksongs, 1903	
24	Ten Pieces, 1894–1903	302
34	Ten Pieces, 1914–16	
40	Pensées lyriques, 1912–14	
41	Kyllikki, 3 nos., 1904	
45/2	Dance Intermezzo, 1904; orchd 1907	302
58	Ten Pieces, 1909	
67	Three Sonatinas, 1912	
68	Two Rondinos, 1912	
—	Spagnuolo, 1913	
74	Four Lyric Pieces, 1914	
75	Five Pieces, 1914	
76	Thirteen Pieces, c1914	
85	Five Pieces, 1916	
—	Mandolinato, 1917	
94	Six Pieces, 1919	
97	Six Bagatelles, 1920	
99	Eight Pieces, 1922	
101	Five Romantic Pieces, 1923	
103	Five Pieces, 1924	
—	Morceau romantique sur un motif de M Jacob de Julin, 1925; orchd as Pièce romantique, 1925, unpubd	
114	Cinq esquisses, 1929, unpubd	

Principal publishers: Breitkopf & Härtel, Hansen, Hirsch, Lienau, Westerlund

Bibliography

BIBLIOGRAPHY

CATALOGUE AND BIBLIOGRAPHY

L. Solanterä: *The Works of Jean Sibelius* (Helsinki, 1955)

F. Blum: *Jean Sibelius: an International Bibliography on the Occasion of the Centennial Celebrations, 1965* (Detroit, 1965)

GENERAL

K. Flodin: *Finska musiker och andra uppsatser i musik* (Helsinki, 1900)

R. Newmarch: *Jean Sibelius: a Finnish Composer* (Leipzig, Brussels, London and New York, 1906)

G. Hauch: *Jean Sibelius* (Copenhagen, 1915)

E. Furuhjelm: *Jean Sibelius: hans tondiktning och drag ur hans liv* (Porvoo [Borgå], 1916)

W. Niemann: *Jean Sibelius* (Leipzig, 1917)

J. H. Elliot: 'Jean Sibelius: a Modern Enigma', *The Chesterian*, vi (1931), 93

C. Gray: *Sibelius* (London, 1931, 2/1945)

G. Bantock: 'Jean Sibelius', *MMR*, lxv (1935), 217

K. Ekman: *Jean Sibelius: en konstnärs liv och personlighet* (Stockholm, 1935, 4/1959; Eng. trans., 1935, 2/1936)

B. Gripenberg: *Till Jean Sibelius på 70-årsdagen* (Helsinki, 1935)

J. Nyyssönen and I. Schiffer: *Jean Sibelius: a nagy finn zeneköltö születésének hetvenedilk évfordulójára kiadta* (Budapest, 1936)

B. de Törne: *Sibelius: a Close-up* (London, 1937)

R. Newmarch: *Jean Sibelius: a Short Story of a Long Friendship* (Boston, 1939, 2/1945)

B. Sandberg: *Jean Sibelius* (Helsinki, 1940)

E. Arnold: *Finlandia: the Story of Sibelius* (New York, 1941, 2/1951)

G. Pirsch: *Jean Sibelius* (Gilly, nr. Charleroi, 1944)

N. Cardus: 'Sibelius', *Ten Composers* (London, 1945, rev. 2/1958)

O. Downes: *Sibelius* (Helsinki, 1945)

M. Similä: *Sibeliana* (Helsinki, 1945)

B. de Törne: *Sibelius i närbild och samtal* (Helsinki and Stockholm, 1945, 2/1955)

G. Abraham, ed.: *Sibelius: a Symposium* (London, 1947, 2/1952)

I. Hannikainen: *Sibelius and the Development of Finnish Music* (London, 1948)

N.-E. Ringbom: *Sibelius* (Stockholm, 1968)

H. Schouwman: *Sibelius* (Haarlem and Antwerp, 1949)

V. Helasvuo: *Sibelius and the Music of Finland* (London, 1952)

——: *Jean Sibelius: a Master and his Work* (Oklahoma, 1954)

O. Andersson: *Jean Sibelius i Amerika* (Turku, 1955)

——: *Jean Sibelius och svenska teatern* (Turku, 1956)

S. Levas: *Jean Sibelius: muistelma suuresta ihmisettä* (Helsinki, 1957, 2/1960; Eng. trans., 1972)

H. Truscott: 'A Sibelian Fallacy', *The Chesterian*, xxxii (1957), 34

A. Forslin: *Runeberg i musiken* (Copenhagen, 1958), 116ff, 205ff, 307ff

H. Johnson: *Jean Sibelius* (New York, 1959, 2/1960)

J. Rosas: 'Sibelius' musik till skådespelet Ödlan', *Smv 1960–61*, 49

R. Simpson: 'Ianus Germanicus: Music in Scandinavia', *Twentieth Century Music*, ed. R. Myers (London, 1960), 165

P. Balogh: *Jean Sibelius* (Budapest, 1961)

E. Tanzberger: *Jean Sibelius: eine Monographie* (Wiesbaden, 1962)

V. Alexandrova and E. Bronfin: *Yan Sibelius: ocherk zhizni i tvorchestva* (Moscow, 1963)

A. M. Stupel: *Yan Sibelius, 1865–1957: kratkiy ocherk zhizni i tvorchestva* (Leningrad, 1963)

M. A. Vachnadze: *Sibelius* (Moscow, 1963)

R. Layton: 'Sibelius – the Early Years', *PRMA*, xci (1964–5), 73

P. Nørgård: 'Sibelius og Danmark', *Smv 1964–5*, 67

D. Schjelderup-Ebbe: 'Sibelius og Norge', *Smv 1964–5*, 80

R. Layton: *Sibelius* (London, 1965, rev. 2/1978)

R. Simpson: *Sibelius and Nielsen* (London, 1965)

M. Vignal: *Jean Sibelius* (Paris, 1965)

E. Tawaststjerna: *Jean Sibelius* (Helsinki, 1965–84; Eng. trans., 1976–)

——: 'Sibelius möte med Debussy i London 1909', *Smv 1967–8*, 31

B. Wallner: *Vår tids musik i Norden* (Stockholm, 1968)

R. Layton: *The World of Sibelius* (London, 1970)

E. Salmenhaara and H. Tirranen: *Sibelius and Ainola* (Porvoo, 1976)

CRITICAL STUDIES

C. Lambert: 'The Symphonies of Sibelius', *The Dominant*, ii/3 (1929), 14

C. Gray: *Sibelius: the Symphonies* (London, 1935)

J. H. Elliott: 'The Sixth Symphony of Sibelius', *ML*, xvii (1936), 234

A. Meyer: 'Sibelius, Symphonist', *MQ*, xxii (1936), 68

A. O. Väisänen: 'Sibelius ja kansanmusükki', *Kalevalaseuran Vuosikirja*, xvi (Porvoo, 1936), 276

E. Roiha: *Die Symphonien von Jean Sibelius: eine formanalytische Studie* (Jyväskylä, 1941)

D. Cherniavsky: 'The Use of Germ Motives by Sibelius', *ML*, xxiii (1942), 1

310

Bibliography

I. Krohn: *Der Formenbau in den Symphonien von Jean Sibelius* (Helsinki, 1942)

N.-E. Ringbom: 'Litteraturen om Jean Sibelius', *STMf*, xxiv (1942), 122

R. Wood: 'Sibelius's Use of Percussion', *ML*, xxiii (1942), 10

E. Tanzberger: *Die symphonischen Dichtungen von Jean Sibelius (eine inhalts- und formanalytische Studie)* (Würzburg, 1943)

C. T. Davie: 'Sibelius's Piano Sonatinas', *Tempo* (1945), no.10

N.-E. Fougstedt: 'Sibelius's tonsättnigar till Rydbergs texter', *Musikvärlden*, i (Stockholm, 1945)

S. von Pfaler: 'Sånger av Sibelius till ord av Runeberg', *Finsk tidskrift*, cxxxviii (1945), 254

I. Krohn: *Der Stimmungsgehalt in den Symphonien von Jean Sibelius* (Helsinki, 1945–6)

E. Tanzberger: *Der Stimmungsgehalt der Symphonien von Jean Sibelius* (Helsinki, 1945–56)

N. V. Bentzon: 'Jean Sibelius och hans nordiske samtid', *Dansk musiktidsskrift*, xxi (1946), 37

J. Jalas: 'Valse triste och musiken till *Kuolema*', *Musikvärlden*, iv (Stockholm, 1948)

N.-E. Ringbom: 'Sibelius och impressionismen', *Finsk tidskrift*, iii (1948), 104

D. Cherniavsky: 'Two Unpublished Tone-poems by Sibelius', *MT*, xc (1949), 272

J. Herbage: 'Jean Sibelius', *The Symphony*, ed. R. Hill (Harmondsworth, 1949)

W. G. Hill: 'Some Aspects of Form in the Symphonies of Sibelius', *MR*, x (1949), 165

W. Mellers: 'Sibelius and the "Modern Mind" ', *Music Survey*, i (1949), 177

D. Cherniavsky: 'Sibelius's Tempo Corrections', *ML*, xxxi (1950), 53

N.-E. Ringbom: 'Sibelius' utvecklingsskeden', *Musikrevy*, v (1950)

E. Tanzberger: 'Jean Sibelius als Symphoniker', *GfMKB, Lüneberg 1950*, 146

R. Gregory: 'Sibelius and the Kalevala', *MMR*, lxxxi (1951), 59

R. Leibowitz: *Jean Sibelius: le plus mauvais compositeur du monde* (Liège, 1955)

S. Parmet: *Sibelius symfonier* (Helsinki, 1955; Eng. trans., 1959)

N.-E. Ringbom: *Sibelius: Symfonier, Symfoniska dikter, Vilkonsert, Voces intimae: analytiska beskrivningar* (Helsinki, 1955)

E. Tawaststjerna: *Sibeliuksen pianosävellykset ja muita esseitä* (Helsinki, 1955; Eng. trans., 1957; Swed. trans., enlarged, 1957)

O. Downes: *Sibelius the Symphonist* (New York, 1956)

311

N.-E. Ringbom: *De två versionerna av Sibelius' tondikt En saga* (Turku, 1956)

J. Rosas: *Otryckta kammarmusikverk av Jean Sibelius* (Turku, 1961)

S. Collins: 'Germ Motives and Guff', *MR*, xxiii (1962), 238

S. Vestdijk: *De symfonieën van Jean Sibelius* (Amsterdam, 1962)

K. Rydman: 'Sibeliuksen neljännen sinfonian rakenncongelmista', *Smv 1962–3*, 17

R. L. Jacobs: 'Sibelius' Lemminkaïnen and the Maidens of Saarii', *MR*, xxiv (1963), 147

Smv 1964–5 [special no.]

O. Ingman: 'Sonaattimuoto Sibeliuksen sinfonioissa', *Smv 1964–5*, 19

J. Rosas: 'Bidrag till kännedom om tre Sibelius-verk', *Smv 1964–5*, 71

B. Wallner: 'Sibelius och den Svenska tonkonsten', *Smv 1964–5*, 91

J. Mattler: 'Quelques aspects de l'être symphonique de Sibelius', *SMz*, cvi (1966)

H. Truscott: 'Jean Sibelius (1865–1957)', *The Symphony*, ii, ed. R. Simpson (Harmondsworth, 1966), 80

L. Normet: 'Reunamerkintöjä Sibeliuksen III sinfoniasta', *Smv 1966–7*, 75

E. Tawaststjerna: 'Jean Sibelius und Claude Debussy (eine Begegnung in London 1909)', *Leoš Janáček et musica europaea: Brno III 1968*, 307

E. Salmenhaara: *Sävelruno Tapiola Sibeliuksen myöhäistyylin edustajana* [The tone poem *Tapiola* as representative of Sibelius's late style] (Helsinki, 1970) [= *Acta musicologica fennica*, iv (1970), 121]

E. Tawaststjerna: 'Sibelius und Bartók: einige Parallelen', *International Musicological Conference in Commemoration of Béla Bartók: Budapest 1971*, 121

——: 'Sibelius' 4. Sinfonie – Schlussphase und Vollendung', *BMw*, xvi (1974), 97

J. Rosas: 'Tondikten Aallottaret (Okeaniderna) opus 73 av Jean Sibelius', *Juhlakirja Erik Tawaststjernalle* (Helsinki, 1976), 37–78

J. Tolonen: 'Jean Sibeliuksen koelunto ja mollipentakordin soinnutus' [The approval lecture by Sibelius and the harmonization of minor pentachords], *Juhlakirja Erik Tawaststjernalle* (Helsinki, 1976), 79

A. Karttunen: 'Tempokäsitysten muutoksista Jean Sibeliuksen sinfonioiden äänilevyesityksissä' [Changes of tempo conceptions in the recording of Sibelius's symphonies], *Juhlakirja Erik Tawaststjernalle* (Helsinki, 1976), 93

F. Tammaro: 'Sibelius e il silenzio di "Tàpiola"', *RIM*, xii (1977), 100–129

Bibliography

L. Pike: *Beethoven, Sibelius and 'The Profound Logic': Studies in Symphonic Analysis* (London, 1978)

B. James: *The Music of Jean Sibelius* (East Brunswick, NJ, and London, 1983)

Index

Text references to works by composers discussed in this volume are indexed in the appropriate work-lists, except for projected or lost works of Mahler and Sibelius listed below.

Index

Index

319

Index

Index

Ueberhorst, Karl, 87
United States of America, 50, 195, 284–5, 290
Universal Edition [music publishers], 14, 16, 25, 48

Vach, Ferdinand, 12
Vaughan Williams, Ralph, 213
Venice, 17
Veni Creator Spiritus, 106, 151
Verdi, Giuseppe, 245
——, *Falstaff*, 99, 236, 240
——, *Il trovatore*, 86
——, *Un ballo in maschera*, 191
Veselá, Marie Calma, 13
Veselý, František, 13
Vestdijk, Simon, 296
Vienna, 5, 6, 7, 14, 81, 82, 86, 87, 96, 99, 100, 101, 104, 106, 107, 111, 187, 195, 197, 198, 203, 207, 230, 233, 235, 272, 275
——, Belvedere, 203, 207
——, Carltheater, 87
——, Conservatory, 5, 82, 84
——, Denkmäler der Tonkunst in Österreich, 101
——, Hofoper, 99, 100, 103, 106, 107
——, Imperial Academy of Music, 284
——, Philharmonic Concerts, 101
——, Philharmonic Orchestra, 99, 272
——, Staatsoper, 202, 210
——, University, 84
——, Vereinigung schaffender Tonkünstler, 104
Vieuxtemps, Henri: *Fantaisie-caprice*, 272
Volga, River, 37
Vyskočil, Q. M., 11

Wagner, Cosima, 192, 193
Wagner, Franziska [wife of A. Ritter], 191
Wagner, Richard, 82, 87, 97, 100, 108, 113, 117, 131, 186, 188, 191, 192, 195, 205, 224, 225, 227, 229, 231, 241, 277, 279, 297
——, *Der Ring*, 92, 100, 186, 194
——, *Das Rheingold*, 91, 92, 95, 185
——, *Die Walküre*, 91, 92, 95, 108
——, *Götterdämmerung*, 96
——, *Siegfried*, 92, 96, 108, 186
——, *Die Meistersinger*, 91, 193, 235, 237, 274
——, *Lohengrin*, 4, 100, 186, 277
——, *Parsifal*, 87, 205
——, *Tannhäuser*, 91, 186, 193, 274
——, *Tristan und Isolde*, 97, 104, 108, 109, 186, 192, 193, 202, 227
Walter, Benno [Franz Joseph Strauss's cousin], 185, 186, 187
Walton, William: Symphony no.1, 303
Warsaw Conservatory, 10
Wasilyeff, Mitrofan, 272
Weber, Carl von [Weber's grandson], 92, 93
Weber, Carl Maria von, 82, 92, 113
——, *Der Freischütz*, 89
——, *Die drei Pintos*, 92–3, 94
Weber, Marian Mathilda von, 93
Webern, Anton, 104, 146
Wegelius, Martin, 272, 274
Weimar, 98, 192, 193, 194, 207
——, Opera, 192
Weingartner, Felix, 194
Welsh National Opera, 50
Wihan, Dora, 218
Wilde, Oscar: *Salome*, 195, 227
Wilde Gung'l [Franz Joseph Strauss's orchestra], 186
Wilhelm II, Kaiser, 221
Wirén, Dag, 303
Wittgenstein, Paul, 203
Wolf, Hugo, 26, 82, 85, 100, 300
——, *Der Corregidor*, 100
Wolzogen, Ernst von, 195
Wood, Henry, 282
Wood, Ralph Walter, 296

323